26 Springer Series in Chemical Physics

Edited by J. Peter Toennies

Springer Series in Chemical Physics
Editors: V. I. Goldanskii   R. Gomer   F. P. Schäfer   J. P. Toennies

B.C. Eu

Semiclassical Theories of Molecular Scattering

With 17 Figures

Springer-Verlag
Berlin Heidelberg New York Tokyo 1984

Professor Byung Chan Eu

McGill University, Department of Chemistry, 801 Sherbrook Street West,
Montreal PQ, Canada H3A 2K6

Series Editors

Professor Vitalii I. Goldanskii

Institute of Chemical Physics
Academy of Sciences
Vorobyevskoye Chaussee 2-b
Moscow V-334, USSR

Professor Robert Gomer

The James Franck Institute
The University of Chicago
5640 Ellis Avenue
Chicago, IL 60637, USA

Professor Dr. Fritz Peter Schäfer

Max-Planck-Institut für
Biophysikalische Chemie
D-3400 Göttingen-Nikolausberg
Fed. Rep. of Germany

Professor Dr. J. Peter Toennies

Max-Planck-Institut für Strömungsforschung
Böttingerstraße 6–8
D-3400 Göttingen
Fed. Rep. of Germany

ISBN-13: 978-3-642-88167-1   e-ISBN-13: 978-3-642-88165-7
DOI: 10.1007/978-3-642-88165-7

Library of Congress Cataloging in Publication Data. Eu, Byung Chan, 1935-. Semiclassical theories of
molecular scattering. (Springer series in chemical physics ; 26). Bibliography: p. 1. Scattering (Physics)
2. Collisions (Physics) 3. Molecular theory. I. Sink, Michael L. II. Title. III. Series: Springer series in
chemical physics ; v. 26. QC794.6.S3E9 1983 539'.6 83-14448

© by Springer-Verlag Berlin Heidelberg 1984
Softcover reprint of the hardcover 1st edition 1984

2153/3130-543210

To my Mother

Editor's Foreword

The study of molecular collisions at energies from less than about 100 eV down to a few 10^{-3} eV, which is roughly the range of chemical interest, has greatly expanded in the last 10 to 20 years. As in many fields, this activity has been stimulated by parallel advances in theory which have triggered the autocatalytic positive feedback system of experiment challenging theory and vice versa. Possibly the biggest driving force, however, has been the growing awareness that molecular collisions are important in our understanding of na- tural and man-made environments. Molecular collision dynamics is now studied in connection with molecular formation in interplanetary space, upper atmo- sphere chemistry, plasmas, lasers and fusion reactors, and is crucial for understanding gas-dynamic flow processes, gas-phase chemical reactions and catalysis. Despite the great strides made in studying elementary collisions in laboratory scattering experiments, many of the processes in these areas are too complicated for us to hope ever to study them in detail in the labo- ratory. Thus in the long run we shall have to rely on theory. Initially, I think many of us, like myself, had hoped that the development of fast compu- ters would outpace the demands on computing time so that "brute force" quan- tum-mechanical exact calculations would provide all the answers. Unfortunate- ly this has not been the case and efficient approximations are needed. They can be broadly classified as classical, semiclassical or semiquantal. Pure classical calculations can, indeed, be useful, but they ignore all the beauti- ful and significant quantum effects and thus the semiclassical approach which combines physical insight with correct quantum-mechanical ingredients is the only salvation. Unfortunately, despite some recent encouraging progress made by M.V. Berry in extending the Ford and Wheeler approach to elastic scattering and by R.A. Marcus and W.H. Miller in treating inelastic and reactive colli- sions, development of semiclassical theories for handling realistic three- dimensional problems has been slow.

In this volume B.C. Eu presents the first comprehensive description of a rigorous semiclassical theory for handling the main collision processes of interest to chemistry: elastic, rotationally inelastic, reactive, curve-crossing, predissociation, and surface-crossing collisions. The monograph starts with a mathematical introduction to some of the problems associated with the nonuniformity of asymtotic expansions and the related Stokes phenomenon. In the next chapter quantum-mechanical scattering theory is developed in the time-dependent and time-independent formalisms. These introductory chapters then form the basis for treating elastic and inelastic scattering. Chapters 7, 8 and 9 form the main part of the monograph, in which the theory is applied to calculations of phase shifts, transition probabilities and S matrices for crossing between two curves, multiple curves and predissociation. The problem of crossing between two potential surfaces, one of the most crucial of the unsolved theoretical problems in collision chemistry, is dealt with next. The book closes with a discussion of scattering from an ellipsoidal particle, a problem of high relevance for inelastic and reactive collisions. Eu has taken great care to develop the theory rigorously and has pointed out the pitfalls and dangers in some currently fashionable theories. Many of the applications of semiclassical theory described here are new or at least have not been presented in a unified form elsewhere.

In my opinion Byung Chan Eu has done both experimentalists and theoreticians in the field of molecular collisions a great service by making this first rigorous comprehensive survey of semiclassical scattering theory available to us. I am confident that this book will serve as a valuable guide in the further development of semiclassical theory and stimulate a deeper, more penetrating insight into the beautiful intricacies of atomic and electronic motion in their fleeting entanglements which we call collisions.

Göttingen, May 1983   *J. Peter Toennies*

Preface

This monograph presents a study of the Wentzel-Kramers-Brillouin (WKB) method and its applications to atomic and molecular scattering. Since semiclassical methods are wide ranging in scope and the WKB method is only one of several semiclassical approximations in quantum mechanics, only a limited number of semiclassical theories are presented here, which I have studied over a number of years. If many worthwhile subjects are omitted, it is not because they are considered unimportant, but because of the limitations on space, time, and energy on my part.

The journey through the realms of semiclassical theories has been a deeply pedagogical and valuable experience for me and it is only appropriate to express my gratitude here to two of my former teachers, Professor Shoon Kyung Kim and Professor John Ross for setting me on the journey and their encouragement. I should like to thank the editor of the Springer Series in Chemical Physics, Professor J. Peter Toennies, for the opportunity to write this monograph. This venture would not have been possible without the valuable help from Dr. Ung-In Cho and Dr. Michael L. Sink, who has done some of the calculations presented. It is my pleasure to take this opportunity to thank them for their assistance. Last but not least, I should like to thank sincerely Mrs. Carol Brown, who has cheerfully and skillfully carried the heavy burden of typing the manuscript under very trying circumstances. I should also like to thank Miss Carla Durston for assisting in the preparation of the manuscript.

Montreal, May 1983   *Byung Chan Eu*

Contents

1. Introduction

The physical chemist is accustomed to studying various rate processes in chemical systems, particularly chemical reactions, on the macroscopic level. To understand the molecular behavior of the systems observed macroscopically, the physical chemist is then compelled to resort to various statistical mechanical methods which tend to average out some interesting and useful information that otherwise might have been observed. This limitation has begun to be lifted since the mid fifties owing to the pioneering work of *Taylor* and *Datz* [1.1] on the application of the molecular-beam technique to chemical reactions, which enabled them to study molecular events. The potential of their work was quickly appreciated by many researchers [1.2] and the field has now grown to a full-fledged discipline which influences thinking in other branches of physical chemistry, such as spectroscopy [1.3] and photochemistry [1.4]. Like other experimental disciplines, molecular-beam studies have created numerous theoretical questions that challenge theoreticians and experimentalists alike, the enormous volume of literature on atomic and molecular scattering theories attesting to the activity in the field.

One most pressing theoretical problem has been, and still is, how to solve the equations of motion—Schrödinger or classical equations—which defy analytic solutions, and calculate cross sections or transition probabilities measured in experiment. There are many ways now known. From the mid sixties theoreticians in the field generally believed that some kind of semiclassical theory, which judiciously blends classical and quantum mechanics, might provide practical and efficient means for studying experimental observations. We now have numerous theories termed "semiclassical" in the literature, indicating the popularity of the term in atomic and molecular scattering theory. One is now obliged to define what is meant by the term. The theories presented in this work are semiclassical in the sense that the equations on which the studies are based are valid in the limit of $\hbar \to 0$, i.e., the classical limit for translational degrees of freedom, while the internal degrees of freedom are generally taken as quantum mechanical. The definition of the term naturally limits the scope of work mainly to the WKB method and related techniques.

The WKB method is an asymptotic expansion theory of differential equations and its basic idea traces back to *Liouville* [1.5]. Being asymptotic, it can provide us with a very good answer in the appropriate range of parameters involved, but it also requires considerable care in application. When coaxed right, it can yield quite satisfactory answers with correct physical pictures added to them as ornaments — something that numerical methods generally do not afford us.

It is in this belief that the semiclassical theories were initially pursued and are presented here. An attempt has been made as far as possible to present them as a coherent whole within the space allotted, but I feel that the potential of the WKB method for molecular scattering still remains largely untapped and further serious studies could lead to higher precision and effectiveness of the method. It is interesting to note that high-energy physicists and field theorists have recently started using the WKB method or variations thereof for their study of quantum chromodynamics [1.6]. This points to the fertility of the idea of the method, which enables us to visualize quantum phenomena in terms of the classical picture, yet still remain in the quantum domain.

Chapter 2 reviews asymptotic expansion methods, particularly Liouville's theory, and the Stokes phenomenon [1.7] which some workers on semiclassical theories tend to overlook especially in computer-assisted trajectory calculations carried out within the semiclassical theory framework. If some of the results were poor in accuracy, it could be quite possible that the fault lay in overlooking the Stokes phenomenon and related aspects, not in the semiclassical approach itself. Out of this belief, an introductory discussion of the phenomenon is given. Chapter 3 defines quantities to be studied and thus fixes some notations. In Chap. 4 elastic scattering is studied with the WKB method mainly by means of uniform semiclassical wave functions. The reader can gain the basic ideas of mathematical machineries used in the later chapters from the discussions given in this chapter. Chapter 5 extends the method introduced in Chap. 4 to inelastic scattering. In Chap. 6 we discuss time-dependent semiclassical theory. This approach seemed very promising, but in application it has not lived up to its expectation in terms of accuracy. Unitary transformations to action-angle variables from variables of continuous spectrum are known to be only one-sided. Some care is required in applying them when semiclassical theories are developed in terms of action-angle variables. It is quite possible that this problem is the source of numerical inaccuracy one sometimes encounters with the method. This aspect is discussed in Chap. 6. Further careful study may make it fully live up to its original expectation. What is presented here is somewhat different from the formalism originally obtained by its proponents. Chapters 7-9 discuss curve-crossing problems, so utilizing the theoretical results obtained in the previous chapters. Chapter 10 attempts to supply a mathematical formalism for classical trajectory calculations for multisurface scattering problems, which may attract those workers interested in surface-hopping processes. Chapter 11 discusses the WKB method for scattering in the prolate spheroidal coordinates which are more appropriate for nonspherical (ellipsoidal) molecules than the usual spherical coordinates commonly used for inelastic scattering studies of nonspherical molecules. There seems to be considerable room for more studies.

2. Mathematical Preparation and Rules of Tracing

Few equations of motion are solvable in terms of known analytic functions in physics and chemistry. We are often compelled to look for some approximate solutions or physical models which yield exactly solvable equations since they furnish more insight into the nature of the system of interest than the exact, but numerical solutions. Frequently, approximations are made on physical rather than on mathematical grounds. Of course, this does not mean that mathematics is completely ignored.

If equations of motion contain one or more parameters which are large compared with other parameters characterizing the physics of the problem, it is quite common to seek solutions in the limit of large parameters identified. The asymptotic solution method is such a technique. It plays an important role in physics and chemical physics and often produces excellent numerical results for the quantities desired, if it is used judiciously and with care. Since this method is employed in this monograph, let us now go over some important features of asymptotic expansions. The material presented is by no means complete, but is aimed at giving the reader the gist of the method used repeatedly throughout this work.

2.1 Asymptotic Expansion

An asymptotic expansion is basically a divergent expansion for a function. An expansion for function $f(z)$

$$f^{(n)}(z) = \sum_{i=0}^{n} f_i z^{-i} \tag{2.1}$$

is called an asymptotic expansion for $f(z)$ in a given range of z if

$$\lim_{|z| \to \infty} |z|^n |f(z) - f^{(n)}(z)| = 0 \tag{2.2}$$

4

$$\lim_{n \to \infty} |z|^n \ |f(z) - f^{(n)}(z)| = 0 \quad , \quad (z \text{ is fixed}) \quad . \tag{2.3}$$

This definition, originally by *Poincaré* [2.1], is generally accepted [2.2].

Despite the inherent divergence of the expansion, (2.1) can be an excellent approximation to $f(z)$ if z is sufficiently large compared with a certain characteristic value. To achieve such an approximation one usually sums the series up to the last diminishing term. The first few terms will generally do for reasonable accuracy in practice. However, being inherently divergent, asymptotic expansions demand care and precaution, unlike convergent expansions, in using the approximate formulas obtained therewith.

Suppose that there is a function $f(\lambda,z)$ of two variables λ and z, which may be expanded into an asymptotic series with respect to z in a given domain of λ and z:

$$f(\lambda,z) = f_0(\lambda) + f_1(\lambda) \ z^{-1} + f_2(\lambda) \ z^{-2} + \ldots \quad . \tag{2.4}$$

This asymptotic expansion generally may not be valid in another domain of variable and therefore is not uniform throughout the whole domain. In fact, since such nonuniformity is generally the rule with an asymptotic expansion considerable caution is required in applying this method in practical computation. It is intimately associated with the Stokes phenomenon [2.3,4] to be discussed shortly below.

There are a number of ways to generate asymptotic expansions. One of them is to utilize an integral representation for the function, e.g., the well-known case of an error function. By integrating the integral by parts repeatedly, an asymptotic series is obtained for the function represented by the integral. Another method is the evaluation of the integral by means of the saddle point or steepest-descent method [2.4-6]. These methods are well-known in classical analysis and have been used for studying the asymptotic expansions of various special functions starting from their integral representations.

Still another method is the Liouville method [2.6,7] which uses differential equations for the function. The method first transforms the latter into a more appropriate form for initiating the asymptotic expansion. This method is often more convenient especially when an integral representation is difficult to find for the function. Since this is predominantly the method we rely on in this monograph, we shall consider some of its basic features. Variations are possible and they will be used later. For the moment we shall consider a typical case.

5

Suppose there is a second-order differential equation

$$\frac{d^2}{dx^2} f(x) + \lambda^2 v(x) f(x) = 0 \quad , \tag{2.5}$$

where λ is a large parameter and $v(x)$ is a continuous differentiable function of x which tends to a limit as $x \to \infty$. We assume the solution of (2.5) exists.

Following *Liouville* [2.7], we introduce the Liouville transformation

$$\xi = \int^x dx \ v^{1/2}(x) \tag{2.6}$$

$$u(x) = v^{1/4}(x) f(x) \quad . \tag{2.7}$$

Then changing the variable to ξ and substituting (2.7) into (2.5), we obtain the differential equation for $u(\xi)$

$$(d^2/d\xi^2)u(\xi) + \lambda^2 u(\xi) = \{v,x\}u(\xi) \quad , \tag{2.8}$$

where

$$\{v,x\} = 4^{-1}(d^2v/dx^2)v^{-2} - (5/16)(dv/dx)^2 v^{-3} \quad , \tag{2.9}$$

which is called the Schwartzian derivative. It is useful to express u in an exponential form

$$u(\xi) = \exp[\lambda S(\lambda,\xi)] \tag{2.10}$$

which on substitution into (2.8) yields a Riccati equation

$$(dS/d\xi)^2 + 1 + \lambda^{-1}(d^2S/d\xi^2) - \lambda^{-2}\{v,x\} = 0 \quad . \tag{2.11}$$

This equation is nonlinear, but can be solved by expanding S in an asymptotic series in λ

$$S = S_0 + S_1\lambda^{-1} + S_2\lambda^{-2} + \dots \quad . \tag{2.12}$$

By substituting (2.12) into (2.11) and equating the coefficients of λ^{-n} to zero, we obtain a hierarchy of equations,

$$(dS_0/d\xi)^2 + 1 = 0 \tag{2.13}$$

$$\frac{d^2S_{n-1}}{d\xi^2} + \sum_{i=0}^{n} \frac{dS_i}{d\xi} \frac{dS_{n-i}}{d\xi} = \{v,x\}\delta_{n2} \quad , \qquad (n = 1,2, \dots) \tag{2.14}$$

where δ_{n2} denotes a Kronecker delta. The hierarchy of equations, when solved successively, yields the solutions as follows:

$$S_0^{(\pm)} = \pm\, i\xi$$

$$\dot{S}_1^{(\pm)} = 0$$

$$S_2^{(\pm)} = \pm\,(i/2) \int^{\xi} d\xi \{v,x\} \tag{2.15}$$

$$S_3^{(\pm)} = 4^{-1}\{v,x\}$$

$$S_4^{(\pm)} = \pm\,(i/8) \int^{\xi} d\xi \left[\frac{d}{d\xi} \{v,x\} - \{v,x\}^2 \right]$$

$$S_5^{(\pm)} = -\,(1/16) \frac{d}{d\xi} \{v,x\} + 8^{-1}\{v,x\}^2$$

... .

It is interesting to note that the solutions for $n \geq 2$ are entirely determined by the Schwartzian derivative and its higher derivatives. Since (2.15)
determines the asymptotic expansions for the phases of the function u, the
asymptotic solution for (2.5) is given by the formula

$$f(x) = v^{-1/4}(x)\Big[A\,\exp(i\lambda\xi - 2^{-1}i\lambda^{-1} \int^{\xi} d\xi \{v,x\} + 4^{-1}\lambda^{-2}\{v,x\} + \ldots)$$

$$+\, B\,\exp(-i\lambda\xi + 2^{-1}i\lambda^{-1} \int^{\xi} d\xi \{v,x\} + 4^{-1}\lambda^{-2}\{v,x\} + \ldots)\Big] \,, \tag{2.16}$$

where A and B are the integration constants. In practice, one rarely takes
terms for $n \geq 2$ in (2.12), since they require more computational effort than
a numerical solution of (2.5) itself. The reader will notice that if only
$S_0^{(\pm)}$ is retained for f(x) the WKB solution to (2.5) is recovered. Histori-
cally, Liouville was the first to develop the present solution technique,
although there are many important contributions [2.8,9] made by numerous
research workers after him and often the method is incorrectly accredited to
later workers.

The asymptotic series thus generated do not converge uniformly and are
not uniformly useful nor meaningful beyond certain regions of parameters.
It is generally necessary to have different asymptotic formulas for f(x) in
different regions. This was not understood until *Stokes* discovered the
phenomenon, now known as the Stokes phenomenon [2.3,4], associated with asymptotic expansions. A proper understanding of this phenomenon is very important and essential for correct usage of the WKB solutions for second-order
differential equations.

2.2 The Stokes Phenomenon

Analytic functions can be expanded in an ascending power series of the argument and their analytic properties can be studied by means of these series. When the argument is small, the ascending power series can be easily used in practice, since a small number of terms are usually sufficient for the required accuracy of the function. However, as the magnitude of the argument gets large, the ascending power series is extremely laborious to use for computation. Besides, it does not even tell us what sort of behavior the function really exhibits in the limit of large argument. To avoid the cumbersome feature of the ascending power series for large argument, it is common to use an asymptotic series for the function, since the asymptotic expansion can give the numerical values of the function much more efficiently if it is used judiciously. However efficient, asymptotic expansions have inherent weaknesses as mentioned before, since they are not only divergent series, but also nonuniform representations of the functions in their entire domain of definition.

To illustrate this point, let us take as an example the function

$$f(z) = 2 \int_0^\infty du \ \exp(-u^2)\sinh(2zu) \tag{2.17}$$

which is a slightly altered form of the function originally taken by Stokes in his pioneering paper on the discontinuities of asymptotic expansions of analytic functions [2.3].

This function is odd with respect to z and analytic. When the hyperbolic function is expanded and the integrals are evaluated term by term, $f(z)$ can be expressed as an ascending series in z:

$$f(z) = \sum_{n=0}^\infty \frac{\Gamma(n + 1)}{(2n + 1)!} (2z)^{2n+1} \quad . \tag{2.18}$$

On the other hand, it obeys the differential equation

$$\frac{df}{dz} - 2zf = 2 \tag{2.19}$$

which may be solved to give

$$f(z) = 2 \ \exp(z^2) \int_0^z du \ \exp(-u^2) \tag{2.20a}$$

if $f(0) = 0$ at $z = 0$. The differential equation is easily obtained from (2.17) by differentiating it. For large positive z we obtain an asymptotic series for $f(z)$ from (2.20a) by integrating by parts repeatedly:

8

$$f(z) = \pi^{1/2} \exp(z^2) - \left(\frac{1}{z} - \frac{1}{2z^3} + \frac{1 \cdot 3}{2^2 z^5} - \frac{1 \cdot 3 \cdot 5}{2^3 z^7} + \ldots \right) \qquad (2.21a)$$

$$\sim \pi^{1/2} \exp(z^2) \quad .$$

This series gives the numerical values of f(z) at large positive z to a high precision. However, the series is no longer odd with respect to z despite the fact that f(z) is odd. If z is negative, then

$$f(-z) = 2 \exp(z^2) \int_0^{-z} du \, \exp(-u^2)$$

$$= - 2 \exp(z^2) \int_0^z du \, \exp(-u^2) \qquad (2.20b)$$

and we then perform integrations by parts to obtain the series

$$f(-z) = - \pi^{1/2} \exp(z^2) + \left(\frac{1}{z} - \frac{1}{2z^3} + \frac{1 \cdot 3}{2^2 z^5} - \frac{1 \cdot 3 \cdot 5}{2^3 z^7} + \ldots \right)$$

$$\sim - \pi^{1/2} \exp(z^2) \quad . \qquad (2.21b)$$

This shows that the coefficients of the asymptotic expansions (2.21a,b) change discontinuously as the argument is changed from one sector to another in the complex z plane. Such discontinuous change in the coefficients in the asymptotic expansion is a universal phenomenon first discovered by *Stokes* [2.3]. A proper understanding of it is essential for correct usage of asymptotic series and any theory based on it. The phenomenon is called the Stokes phenomenon after its discoverer and the correspondences of different asymptotic formulas in different regions are called the connection formulas. Since the WKB (semiclassical) method is essentially an asymptotic solution method, it is easy to see that the Stokes phenomenon plays an important role in semiclassical theories based on the WKB method of solution [2.10-12]. Its inappropriate application can lead to absurd results in quantum mechanics. In fact, it plays a crucial role in the development of semiclassical theories for inelastic scattering phenomena, as will be seen in later chapters. For this reason, we shall elaborate on it further with a more practical and important example, the Airy functions [2.13], which form the basis of many discussions in later chapters in this monograph.

2.3 The Stokes Phenomenon Continued: The Airy Functions

Since they are the simplest nontrivial functions with one turning point, Airy functions are most extensively studied in connection with asymptotic expansions. The Airy functions Ai(z) and Bi(z) may be defined by the integrals [2.5,6,13,14],

$$Ai(z) = (2\pi i)^{-1} \int_C dt \, \exp(tz - t^3/3) \qquad (2.22a)$$

$$Bi(z) = (2\pi i)^{-1} \left(\int_{C_1} dt + \int_{C_2} dt \right) \exp(tz - t^3/3) \ , \qquad (2.22b)$$

where the contours are depicted in Fig.2.1. On differentiating (2.22) twice with respect to z, we find the differential equation for them

$$(d^2/dz^2)f(z) - zf(z) = 0 \ , \qquad (2.23)$$

where f is Ai or Bi. In fact, Ai(z) and Bi(z) are two linearly independent solutions of (2.23).

Since the analysis presented below applies equally well to both functions we shall carry it out in terms of Ai(z) and list the results for Bi(z) without detailed analysis.

By change of variables

$$t = r^{1/2}u \ , \quad z = r \, \exp(i\theta) \ ,$$

the integral representation for Ai(z) may be cast into the form

$$Ai(z) = r^{1/2}I(r,\theta) \ , \qquad (2.24a)$$

where

$$I(r,\theta) = (2\pi i)^{-1} \int_C du \, \exp\{\zeta[u \, \exp(i\theta) - u^3/3]\} \qquad (2.24b)$$

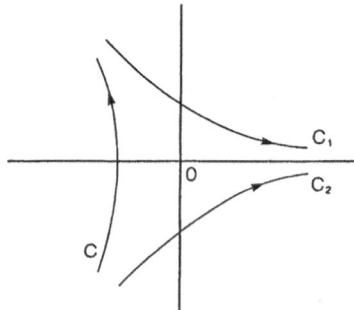

Fig.2.1. Integration paths of Airy functions

with

$$\zeta = r^{3/2} > 0 \quad .$$

For large ζ, i.e., large r, the integral may be evaluated asymptotically by using the method of steepest descent. Since the saddle points are located at

$$u_0 = \pm \exp(i\theta/2) \quad ,$$

the integral may be written as

$$I = \exp(\pm\xi)I^{(\pm)} \quad ,$$

where

$$I^{(\pm)} = (2\pi i)^{-1} \int_C ds \, \exp[\pm\zeta \, \exp(i\theta/2)s^2 - \zeta s^3/3] \tag{2.25}$$

$$\xi = (2/3)r^{3/2} \exp(i3\theta/2) = (2/3)\zeta^{3/2} \exp(i3\theta/2) = 2z^{3/2}/3$$

$$s = u - u_0 \quad .$$

The integrals in (2.25) are evaluated asymptotically. Different sectors of θ require separate treatments.

If $-\pi/3 < \theta < \pi/3$, the saddle points are generally complex and as $\theta \to 0$, they tend to ± 1. When $\theta = 0$, the second derivative of the phase function in (2.25) is real and positive at the negative saddle point and therefore the steepest descent becomes parallel to the imaginary axis. As the ray moves away from the positive real axis in either direction, the steepest descent becomes parallel to the rays $\phi = \pm\pi/2 - \theta/4$ issuing from the negative (-1) saddle point. Therefore, by distorting the contour to the rays, we obtain from (2.25) the following integral,[1]

$$I^{(-)} = (1/2\pi i) \left(\int_{\infty \, \exp(-i\pi/2-i\theta/4)}^{0} + \int_{0}^{\infty \, \exp(i\pi/2-i\theta/4)} \right)$$

$$ds \, \exp\{\zeta[s^2 \exp(i\theta/2) - s^3/3]\}$$

which, after a change of variable, takes the form

$$I^{(-)} = (2\pi)^{-1}\zeta^{-1/2} \exp(-i\theta/4) \int_{0}^{\infty} dy \, y^{-1/2} \exp(-y)\cos[(3\xi/2)^{-1/2}y^{3/2}/3] \quad .$$

---

[1] The end points $\infty\exp(\pm i\pi^{1/2} - i\theta^{1/4})$ below and in all subsequent integrals mean that the integrations are to be performed along the rays issuing from the origin with the phase angles $\pm\pi^{1/2} - \theta^{1/4}$ in the complex plane.

When the cosine function is expanded in the Taylor series, each term can be evaluated easily and we finally obtain an asymptotic series for the integral:

$$I^{(-)} = (2\pi)^{-1}\zeta^{-1/2} \exp(-i\theta/4) \sum_{\ell=0}^{\infty} \frac{\Gamma(6\ell + 1/2)}{(2\ell)!3^{2\ell}} (-3\xi/2)^{-\ell} \quad . \tag{2.26}$$

As a matter of fact, this formula remains valid when θ is extended up to $+2\pi/3$.

Substituting this result into (2.24) and expressing the series in terms of z yields the asymptotic form for Ai(z):

$$Ai(z) = 2^{-1}\pi^{-1/2}z^{-1/4} \exp\left(-\frac{2}{3} z^{3/2}\right) \sum_{\ell=0}^{\infty} \frac{(-)^{\ell}\Gamma(6\ell + 1/2)}{\Gamma(1/2)\Gamma(2\ell + 1)3^{2\ell}} z^{-3\ell/2} \quad ,$$

$$(- 2\pi/3 < \theta < 2\pi/3) \quad . \tag{2.27}$$

To see the behavior of Ai(z) in the sector $2\pi/3 < \theta < 4\pi/3$, we consider the case of $\theta = \pi$ which means that the saddle points are now located at $\pm i$. Then the steepest descent is along the ray $\theta = \pi/4$ for the lower saddle point and along the ray $\theta = -\pi/4$ for the upper saddle point. Therefore, the contour must be distorted, as indicated in Fig.2.2. Then we have

$$I^{(\pm)} = (2\pi i)^{-1}\left(\mp \int_{\pm i\ \exp(+i\pi/4)}^{\infty} \pm \int_{\pm\exp(\pm i3\pi/4)}^{\infty\ \exp(\pm i3\pi/4)}\right)ds\ \exp[\mp i\zeta(s\mp i)^2 - \zeta(s\mp i)^3/3] \quad .$$

The first integral may be distorted into the rays $\theta = \pm\pi/4$ without changing the value of the integral. Thus we finally obtain

$$I^{(\pm)} = \pi^{-1} \exp(\pm i\pi/4) \int_{0}^{\infty} dR\ \exp(-\zeta R^2)\cos[\zeta\ \exp(\mp i\pi/4)R^3/3]$$

$$= \frac{\exp(\pm i\pi/4)}{2\pi\zeta^{1/2}} \sum_{\ell=0}^{\infty} \frac{(\pm 2)^{\ell}\Gamma(3\ell + 1/2)}{\Gamma(2\ell + 1)3^{3\ell}} (2\zeta/3)^{-\ell} \quad . \tag{2.28}$$

Combining this result with (2.24), we obtain the Airy function in the sector $2\pi/3 < \theta < 4\pi/3$:

$$Ai(-z) = \pi^{-1/2}z^{-1/4}\left[\sin(2z^{3/2}/3 + \pi/4) \sum_{\ell=0}^{\infty} \frac{(-)^{\ell}\Gamma(6\ell + 1/2)}{54^{2\ell}\Gamma(2\ell + 1)\Gamma(2\ell + 1/2)} (2\zeta/3)^{-2\ell}\right.$$

$$\left. - \cos(2z^{3/2}/3 + \pi/4) \sum_{\ell=0}^{\infty} \frac{(-)^{\ell}\Gamma(6\ell + 7/2)}{54^{2\ell+1}\Gamma(2\ell + 2)\Gamma(2\ell + 3/2)} (2\zeta/3)^{-2\ell-1}\right]$$

$$\sim \pi^{-1/2}z^{-1/4} \sin(2z^{3/2}/3 + \pi/4) \quad . \tag{2.29}$$

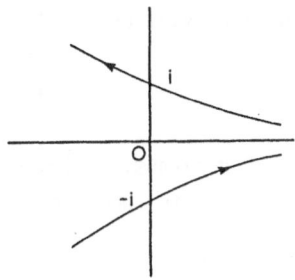
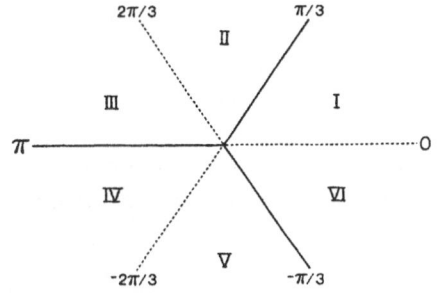

Fig.2.2. Integration paths for Airy functions

Fig.2.3. Stokes and anti-Stokes lines. The solid lines are the anti-Stokes lines, the dotted, the Stokes lines

In summary, we indicate different asymptotic formulas in different sectors in Fig.2.3:

I, VI: $Ai(z) \sim 2^{-1} \pi^{-1/2} z^{-1/4} \exp(-2z^{3/2}/3)$ (2.30a)

II, V: $Ai(z) \sim 2^{-1} \pi^{-1/2} z^{-1/4} \exp(-2z^{3/2}/3)$ (2.30b)

III, IV: $Ai(-z) \sim \pi^{-1/2} z^{-1/4} \sin(2z^{3/2}/3 + \pi/4)$.   (2.30c)

These concisely show that the coefficients in the asymptotic expansion change discontinuously as the argument of the variable changes from 0 to 2π and so on.

Applying similar methods of evaluation to the integral representation for Bi(z), we also find different asymptotic formulas in different sectors. The results are summarized as follows:

I, VI: $Bi(z) \sim \pi^{-1/2} z^{-1/4} \exp(2z^{3/2}/3)$ (2.31a)

II, V: $Bi(z) \sim \pi^{-1/2} z^{-1/4} \exp(2z^{3/2}/3)$ (2.31b)

III, IV: $Bi(-z) \sim \pi^{-1/2} z^{-1/4} \cos(2z^{3/2}/3 + \pi/4)$.   (2.31c)

2.4 The Rules of Tracing

Since the changes in the asymptotic formulas for Ai(z) and Bi(z) just discussed provide very useful tracing rules applicable to single-turning-point problems and, with slight modification, to two-turning-point problems (to be considered later), we shall investigate them in more detail here.

The rays $\theta = \pm\pi/3$, π are called the anti-Stokes lines along which $Re\{z^{3/2}\} = 0$ and the rays $\theta = \pm2\pi/3$, 0 the Stokes lines along which

$\text{Im}\{z^{3/2}\} = 0$. It is convenient to define the following abbreviations [2.15]

$$(0,z) = z^{-1/4} \exp(2z^{3/2}/3)$$

$$(z,0) = z^{-1/4} \exp(-2z^{3/2}/3) \quad .$$

(2.32)

Then in I and VI $(z,0)$ is subdominant, while $(0,z)$ is dominant. To indicate dominancy and subdominancy, we affix subscripts s and d to them. For example, in I and VI

$$(z,0) \rightarrow (z,0)_s$$

$$(0,z) \rightarrow (0,z)_d \quad .$$

With these notations we may write $Ai(z)$ in I and VI as

$$Ai(z) = c^{(+)}(0,z)_d + c^{(-)}(z,0)_s \quad ,$$

(2.33)

where $c^{(+)} = 0$ and $c^{(-)} = 1/2\pi^{1/2}$ according to the analysis leading to (2.30). In the language of asymptotic expansions the coefficient of $(0,z)_d$ is negligibly small compared to that of $(z,0)_s$ in the sectors considered. Nevertheless, it is useful to keep it for the sake of discussion. Now, as the argument of z is increased, $(z,0)_d$ remains subdominant until $\theta = \pi/3$ is reached and as the anti-Stokes line at $\theta = \pi/3$ is crossed $(z,0)_s$ turns into the dominant term, while $(0,z)_d$ becomes subdominant. Thus $Ai(z)$ may be written as

$$Ai(z) = c^{(+)}(0,z)_s + c^{(-)}(z,0)_d \quad .$$

(2.34)

In other words, the dominancy and subdominancy of $(0,z)$ and $(z,0)$ are interchanged as the anti-Stokes line is crossed. The newly established dominancy relation does not change until the next anti-Stokes line at $\theta = \pi$ is crossed. Note that (2.34) is the same as (2.30b), when put back into its more conventional form and if the subdominant term is neglected.

Now the function $Ai(-z)$ in III and IV may be written as

$$Ai(-z) = c^{(+)} \exp(-i\pi/4)(0,z)_s + c^{(-)} \exp(i\pi/4)(z,0)_d$$

(2.35)

where

$$c^{(\pm)} = (2\pi)^{-1/2} \quad .$$

The subscripts s and d should be dropped on the ray $\theta = \pi$, since the distinction cannot be clearly made between them anymore in the sector. Thus at $\theta = \pi$ we write

$$Ai(-r) = c^{(+)} \exp(i\pi/4)(0,r) + c^{(-)} \exp(-i\pi/4)(r,0)$$

where

$$(0,r) = r^{-1/2} \exp(-i2r^{3/2}/3)$$

$$(r,0) = r^{-1/2} \exp(i2r^{3/2}/3) \quad .$$

Comparing (2.34,35), we now see that as the Stokes line at $\theta = 2\pi/3$ is crossed, the coefficient $C^{(+)}$ of the subdominant term in (2.34), which has been negligible up to now, has gone through a discontinuous change to a non-zero value yielding the asymptotic form (2.30c) in Sectors III and IV. Such a discontinuous change in coefficients must happen for the analytic function approximated by a nonanalytic asymptotic series to preserve its integrity in the whole domain of the variable. It is another example of the Stokes phenomenon which *Stokes* himself discussed in great detail [2.3]. The discussion proceeds similarly for tracing Ai(z) in the lower half plane.

The precise forms for asymptotic expansions of a function in different sectors can be obtained by carefully investigating the integral representations of the function in different sectors if the integral representation is known. Unfortunately, since integral representations are not always known for differential equations obeyed by wave functions for systems of practical interest, the usual procedure is to reduce the differential equations in the vicinity of the turning points to that of an Airy function, and then use the asymptotic formulas of the Airy function locally to deduce appropriate forms of the asymptotic expansion for the function of interest. This program was originally carried out by *Zwaan* [2.16], *Kemble* [2.17] and *Langer* [2.18].

Their analyses also confirm the rule concerning the discontinuous change in the coefficient of the subdominant term in the asymptotic solutions for differential equations with one or two turning points. Based on these results, *Heading* [2.15] reduced the tracings of asymptotic forms for some functions in different sectors in the complex plane to a set of rules. Since these rules turn out to be applicable to some inelastic scattering problems and therefore of general interest in this monograph, they are summarized below.

Since asymptotic forms for a function are generally multi-valued functions, they must be made single-valued by introducing branch cuts in the complex plane. As the asymptotic forms are traced across the branch cut, they change discontinuously. The following rules can be easily proved.[2]

---

2 The rules have been reproduced from [2.15] with the author's permission.

Rule 1. As asymptotic forms are traced positively (counter-clockwise) across the branch cut, they transform as follows:

$$(0,z) \rightarrow -i(z,0)$$

$$(z,0) \rightarrow -i(0,z) \quad , \tag{R.1}$$

where $(0,z)$ and $(z,0)$ are defined by (2.32). Dominancy or subdominancy is preserved in the process. If they are traced negatively (clockwise), $-i$ is replaced by $+i$.

It was shown that the coefficient of the subdominant term changes discontinuously as a Stokes line is crossed. Since the precise value of change depends in general on the location of the Stokes line, we associate a constant, called the Stokes constant, with each Stokes line. The following rule holds for the coefficient of a subdominant term.

Rule 2. As a subdominant term is traced positively across a Stokes line, then

new subdominant coefficient = old subdominant coefficient
$$+ T \times \text{dominant coefficient} \quad , \tag{R.2}$$

where T is the Stokes constant for the Stokes line. The dominant coefficient remains unchanged.

This set of rules gives rise to a set of algebraic equations for the Stokes constants, as asymptotic forms are traced around, and their solution yields the Stokes constants. With Stokes constants thus determined, one can obtain the connections of various asymptotic formulas for the function in different sectors in the complex plane. *Heading* [2.15] showed that these rules uniquely determine the Stokes constants for Airy functions or single-turning-point problems, but give rise to an underdetermined system of equations for Stokes constants in the case of two-turning-point problems. However, in the latter case it is possible to determine them uniquely by investigating the asymptotic forms in more detail by means of an integral representation of the parabolic cylinder function that is the reference function. Once thus determined, they can be used for other problems in the same category. These rules are very useful for developing qualitative theories of inelastic scattering processes.

3. Scattering Theory of Atoms and Molecules

The quantum-mechanical description of collision processes between particles such as molecules, atoms, etc., can be approached in two basically equivalent ways if the forces are conservative. One is the time-dependent theory and the other is the time-independent theory. If the equations of motion involved are solved exactly, it does not matter which approach is taken to calculate transition probabilities, but since one approximation method or another must be employed in practice, one approach is more advantageous than the other for a given type of approximation.

Since both approaches are considered here for developing semiclassical approximations, they are briefly discussed to define notations and present equations on which semiclassical approximations will be based later. The aim of this chapter is not to develop scattering theory. For fuller accounts of scattering theory the reader is referred to [3.1-6].

3.1 Time-Dependent Scattering Theory

Suppose a structureless projectile is in collision with another structureless particle. If the interaction force is finite ranged, then at the initial time when the two particles are infinitely separated, the Hamiltonian of the system in the center of mass coordinate system at rest is simply the kinetic energy of the relative motion

$$H_0 = p^2/2m \quad , \tag{3.1}$$

where p is the relative momentum and m the reduced mass. As the particles move toward each other, the strength of interaction increases until it reaches the full value at the closest approach and the Hamiltonian may be written as

$$H = H_0 + H_1 \quad , \tag{3.2}$$

where H_1 is the interaction Hamiltonian (energy). As the particles recede from each other, H_1 diminishes in magnitude and eventually vanishes. Then the collision is completed between them. This picture can be translated into quantum-mechanical language more appropriate for our purpose by using the concept of wave packets. We shall be using an equivalent, but formal description here.

The Schrödinger equation is

$$i\hbar \frac{\partial \Psi}{\partial t} = H\Psi \quad . \tag{3.3}$$

In the remote past the interaction is negligible and the wave evolves according to

$$i\hbar \frac{\partial \Phi}{\partial t} = H_0\Phi \quad . \tag{3.4}$$

Formal solution of (3.3,4) can be facilitated by introducing Green's functions satisfying the inhomogeneous differential equations

$$\left(i\hbar \frac{\partial}{\partial t} - H_0\right)G_0^{(\pm)}(t) = 1\delta(t) \tag{3.5}$$

$$\left(i\hbar \frac{\partial}{\partial t} - H\right)G^{(\pm)}(t) = 1\delta(t) \quad . \tag{3.6}$$

The Green functions are subject to the following initial conditions:

$$G_0^{(+)}(t) = G^{(+)}(t) = 0 \quad , \quad (t < 0)$$
$$G_0^{(-)}(t) = G^{(-)}(t) = 0 \quad , \quad (t > 0) \quad . \tag{3.7}$$

Notice that the + and - solutions are not continuous at $t = 0$. This discontinuity is similar to that of the Green function for second-order differential operators in the theory of differential equations [3.7-10]. Then the formal solutions of (3.5,6) subject to condition (3.7) are

$$G_0^{(+)}(t) = \begin{cases} -i \exp(-i\hbar^{-1}H_0 t) & t > 0 \\ 0 & t < 0 \end{cases} \tag{3.8a}$$

$$G_0^{(-)}(t) = \begin{cases} 0 & t > 0 \\ i \exp(-i\hbar^{-1}H_0 t) & t < 0 \end{cases} \tag{3.8b}$$

$$G^{(+)}(t) = \begin{cases} -i \exp(-i\hbar^{-1}H t) & t > 0 \\ 0 & t < 0 \end{cases} \tag{3.9a}$$

18

$$G^{(-)}(t) = \begin{cases} 0 & t > 0 \\ i \exp(-i\hbar^{-1}Ht) & t < 0 \end{cases} .$$

(3.9b)

These formal solutions will be useful for later discussions. It is possible
to obtain another form of formal solutions. Split H in (3.6) into H_0 and H_1
according to (3.2) and rearrange the H_1 term. When the equation is then
multiplied with $G_0^{(\pm)}$ and integrated over the entire range of time, there
follows the integral equation for $G^{(\pm)}$

$$G^{(\pm)}(t) = G_0^{(\pm)}(t) + \int_{-\infty}^{\infty} dt'\, G^{(\pm)}(t - t')H_1 G_0^{(\pm)}(t') ,$$

(3.10a)

or alternatively

$$G^{(\pm)}(t) = G_0^{(\pm)}(t) + \int_{-\infty}^{\infty} dt'\, G_0^{(\pm)}(t - t')H_1 G^{(\pm)}(t')$$

(3.10b)

where $t > t'$. The wave functions may then be expressed in terms of the Green
functions. For $t > t'$

$$\Phi(t) = iG_0^{(+)}(t - t')\Phi(t')$$

$$\Psi(t) = iG^{(+)}(t - t')\Psi(t')$$

(3.11)

and for $t < t'$

$$\Phi(t) = -iG_0^{(-)}(t - t')\Phi(t')$$

$$\Psi(t) = -iG^{(-)}(t - t')\Psi(t') .$$

(3.12)

Since the Green functions obey the integral equations, (3.11,12) suggest
that the wave functions can also be given in equivalent integral equations,
provided that limits of the wave function $\Psi(t)$ exist in the remote past and
distant future:

$$\Psi^{(+)}(t) = \Phi^{(in)}(t) + \int_{-\infty}^{\infty} dt'\, G_0^{(+)}(t - t')H_1\Psi^{(+)}(t')$$

(3.13a)

$$\Psi^{(-)}(t) = \Phi^{(out)}(t) + \int_{-\infty}^{\infty} dt'\, G_0^{(-)}(t - t')H_1\Psi^{(-)}(t') ,$$

(3.13b)

where "in" and "out" states are defined by the limits

$$\Phi^{(in)}(t) = \lim_{t' \to -\infty} iG_0^{(+)}(t - t')\Psi(t')$$

(3.14a)

$$\Phi^{(out)}(t) = \lim_{t' \to \infty} -iG_0^{(-)}(t - t')\Psi(t') .$$

(3.14b)

The existence of these limits is an important, basic premise of scattering theory. In view of (3.3,8,9) these limits may be expressed formally as

$$\lim_{t \to \infty, t' \to -\infty} \exp[-i\hbar^{-1}(t - t')H_0]\exp[-i\hbar^{-1}(t - t')H] \quad . \tag{3.14c}$$

The existence of this limit is essential for transition probabilities to exist as will be seen shortly.

The Møller wave operator is defined by [3.11]

$$\psi^{(+)}(t) = \Omega^{(+)}\phi^{(in)}(t) \tag{3.15}$$

which, when compared with (3.13a), yields $\Omega^{(+)}$ in the form

$$\Omega^{(+)} = 1 - i \int_{-\infty}^{\infty} dt' \, G^{(+)}(t - t')H_1 G_0^{(-)}(t - t') \quad .$$

By shifting the time and using (3.3,8,9), we may integrate to obtain the operator in the form

$$\Omega^{(+)} = \lim_{t \to -\infty} \exp(i\hbar^{-1}Ht)\exp(-i\hbar^{-1}H_0 t) \quad . \tag{3.16a}$$

Since $\exp(i\hbar^{-1}Ht)$ and $\exp(-i\hbar^{-1}H_0 t)$ are strongly continuous functions of t, their product $\Omega^{(+)}(t)$ is also strongly continuous. Consequently, for functions ϕ_1, $\phi_2 \in H$ (Hilbert space)

$$\langle \phi_1 | \Omega^{(+)}(t)\phi_2 \rangle = \langle \phi_1 | \exp(i\hbar^{-1}Ht) \, \exp(-i\hbar^{-1}H_0 t)\phi_2 \rangle$$

is continuous and bounded in t. Then it can be shown [3.5] that a unique bounded operator $\Omega^{(+)}(\varepsilon)$ exists such that

$$\langle \phi_1 | \Omega^{(+)}(\varepsilon)\phi_2 \rangle = \varepsilon \int_{-\infty}^{0} dt \, e^{\varepsilon t} \langle \phi_1 | \exp(i\hbar^{-1}Ht) \, \exp(-i\hbar^{-1}H_0 t)\phi_2 \rangle$$

for any $\varepsilon > 0$ and for all ϕ_1, $\phi_2 \in H$. This relation is usually written in the operator form

$$\Omega^{(+)}(\varepsilon) = \varepsilon \int_{-\infty}^{0} dt \, e^{\varepsilon t} \, \exp(i\hbar^{-1}Ht) \, \exp(-i\hbar^{-1}H_0 t) \quad . \tag{3.16b}$$

Since the operator is bounded, then

$$\Omega^{(+)} = \lim_{\varepsilon \to +0} \Omega^{(+)}(\varepsilon) \tag{3.16c}$$

for an alternative form of the wave operator. Similarly, the other wave operator is defined by

$$\psi^{(-)}(t) = \Omega^{(-)}\phi^{(out)}(t) \quad , \tag{3.17}$$

which yields

$$\Omega^{(-)} = 1 + i \int_{-\infty}^{\infty} dt' \ G^{(-)}(t - t')H_1 G_0^{(+)}(t' - t)$$

$$= \lim_{t \to \infty} \exp(i\hbar^{-1}Ht) \ \exp(-i\hbar^{-1}H_0 t)$$

$$= \lim_{\epsilon \to +0} \epsilon \int_0^{\infty} dt \ e^{-\epsilon t} \ \exp(i\hbar^{-1}Ht) \ \exp(-i\hbar^{-1}H_0 t) \quad . \tag{3.18}$$

The scattering matrix is then defined as the operator that transforms the "in" state $\phi^{(in)}(-\infty)$ to the "out" state $\phi^{(out)}(\infty)$

$$\phi^{(out)}(\infty) = S\phi^{(in)}(-\infty) \quad . \tag{3.19}$$

From (3.15,17) it then follows that

$$S = \Omega^{(-)+}\Omega^{(+)} \tag{3.20}$$

$$= \lim_{t_0 \to -\infty, t \to \infty} \exp(i\hbar^{-1}H_0 t) \ \exp[-i\hbar^{-1}H(t - t_0)] \ \exp(-i\hbar^{-1}H_0 t_0) \quad .$$

In time-dependent scattering theory one attempts to calculate (3.20) in one way or another. This form of the S matrix is the starting point for a semiclassical approximation method presented later.

The formalism presented above holds for the situation where only elastic or inelastic scattering is possible. If rearrangements are possible between colliding particles, then the formalism requires appropriate revision. Since chemical reactions are important examples of rearrangement collisions, rearrangement scattering theory has drawn considerable attention. What is presented here is the essence of the difference between rearrangement collision theory and nonrearrangement theory and of what is needed to construct semiclassical approximation theories.

The complications of rearrangement theories basically stem from the non-existence of the limits such as (3.14c,20), if rearrangements are allowed to happen between colliding partners, since the asymptotic states can then be basically different before and after a rearrangement.

Let us define arrangement channels as the states of m groups of bound particles formed from n particles participating in the collision [3.3]. The Hamiltonian of a group is denoted by $H_a^{(i)}$, where i refers to the particular group in the arrangement channel a. Then the arrangement channel Hamiltonian is given as the sum of the group Hamiltonians

$$H_a = \sum_{i=1}^{m} H_a^{(i)} \quad . \tag{3.21}$$

The eigenfunctions for H_a in the space $\mathcal{H}_a$ of arrangement channel a are not orthogonal to those in space $\mathcal{H}_b$ of arrangement channel b if a $\neq$ b. This non-orthogonality is another aspect of the reason for the nonexistence of (3.20), for example.

In view of the possibility of rearrangements into different groups of particles the total Hamiltonian can be partitioned differently,

$$H = H_a + H_a' = H_b + H_b' = \ldots \quad , \tag{3.22}$$

where H_a' represents the interaction energy between the m groups in arrangement channel a, and similarly for H_b', etc. The chemical reactions

$$C_6H_5Cl + F \rightarrow \left\{ \begin{array}{l} C_6H_4ClF + H \\[2ex] C_6H_5F + Cl \end{array} \right.$$

exemplify the partitioning (3.22). In this example, there are three rearrangement channels a = (C_6H_5Cl, F), b = (C_6H_4ClF, H), c = (C_6H_5F, Cl) and H_a is the sum of the Hamiltonians of completely isolated pairs in channel a, and similarly for H_b and H_c.

Since the interaction vanishes at infinite separation of the groups (fragments) in a channel, we define the Green function $G_a^{(\pm)}$ by

$$\left(i\hbar - \frac{\partial}{\partial t} - H_a \right) G_a^{(\pm)}(t) = 1\delta(t) \quad , \tag{3.23}$$

which generalizes (3.3). The formal solutions subject to appropriate initial conditions are

$$G_a^{(+)}(t) = \left\{ \begin{array}{ll} -i \, \exp(-i\hbar^{-1}H_a t) & t > 0 \\[2ex] 0 & t < 0 \end{array} \right. \tag{3.24a}$$

$$G_a^{(-)}(t) = \left\{ \begin{array}{ll} 0 & t > 0 \\[2ex] i \, \exp(-i\hbar^{-1}H_a t) & t < 0 \end{array} \right. \tag{3.24b}$$

With these Green's functions we now generalize the "in" and "out" states

$$\phi_\alpha^{(in)}(t) = \lim_{t' \to -\infty} iG_a^{(+)}(t - t')\psi_\alpha^{(+)}(t) \tag{3.25a}$$

$$\phi_\alpha^{(out)}(t) = \lim_{t' \to \infty} -iG_a^{(-)}(t - t')\psi_\alpha^{(-)}(t) \quad . \tag{3.25b}$$

Then by making use of these equations and the integral equation for $G^{(\pm)}$

$$G^{(\pm)}(t) = G_a^{(\pm)}(t) + \int_{-\infty}^{\infty} dt' G_a^{(\pm)}(t - t')H_a'G^{(\pm)}(t')$$

$$= G_a^{(\pm)}(t) + \int_{-\infty}^{\infty} dt' G^{(\pm)}(t - t')H_a'G_a^{(\pm)}(t') \quad , \qquad (3.26)$$

we obtain the integral equations for the scattering wave functions $\psi_\alpha^{(\pm)}(t)$:

$$\psi_\alpha^{(\pm)}(t) = \Psi_{a\alpha}(t) + \int_{-\infty}^{\infty} dt' G_a^{(\pm)}(t - t')H_a'\psi_\alpha^{(\pm)}(t) \quad , \qquad (3.27)$$

where

$$\Psi_{a\alpha}(t) = \phi_\alpha^{(in)}(t) \qquad\qquad \text{for } (+)$$

$$\Psi_{a\alpha}(t) = \phi_\alpha^{(out)}(t) \qquad\qquad \text{for } (-) \quad .$$

The label a stands for the arrangement channel as well as the quantum numbers necessary to specify the state.

The theory of nonreactive scattering discussed before, and the definition of Møller's wave operator in particular, is centered on the existence of the limit (3.14). However, if the Hamiltonian H can be partitioned in a number of different ways as indicated in (3.22) and if H_a', H_b', etc., become negligible at remote past and distant future, then the operator

$$\exp[-i\hbar(t - t')H_a]\exp[i\hbar(t - t')H]$$

no longer has a limit as $t \to \infty$ and $t' \to -\infty$.

This difficulty of the nonexistence of limits for some operators in the case of reactions is eliminated if orthogonal projections P_a onto the channel space H_a are introduced in the formulation by following *Ekstein* [3.12], *Faddeev* [3.13] and others [3.3,5]. We define projection operators P_a such that

$$P_a^2 = P_a \quad , \quad P_aH_a = H_a \quad . \qquad\qquad (3.28)$$

The null space of P_a is spanned by the states of completely dissociated fragments in channel a, i.e., the continuous spectrum.

We are now able to generalize the Møller wave operators with the help of the projection operators just introduced. The generalization proceeds as follows. First let us observe that as $t \to \pm\infty$

$$\psi^{(\pm)}(t) \to \sum_\alpha \exp[-i\hbar^{-1}H_a(t - t')]\Psi_{a\alpha}(t') \quad .$$

Therefore operating P_a on the wave function, we find

$$P_a\psi^{(\pm)}(t) \to \exp[-i\hbar^{-1}H_a(t - t')]\Psi_{a\alpha}(t') \quad .$$

The right-hand side is well defined and has a limit as $t \to \infty$. Thus if we define operator $\Omega_a(t - t')$ by

$$\Omega_a(t - t') = \exp[i\hbar^{-1}H_a(t - t')]P_a \exp[-i\hbar^{-1}H(t - t')] \quad ,$$

then the limit exists. We therefore define the Møller operators by

$$\psi_\alpha^{(\pm)}(t) = \Omega_a^{(\pm)}\psi_{a\alpha}(t') \quad , \tag{3.29}$$

where the superscript $(\pm)$ on the wave operator distinguishes the "in" and "out" states. Comparing (3.29) with (3.27) together with (3.25) yields an integral form for the Møller wave operator,

$$\Omega_a^{(+)} = 1 - i \int_{-\infty}^{\infty} dt \; G^{(+)}(-t)H_a'G_a^{(-)}(t)P_a$$

$$= \lim_{t \to -\infty} \exp(i\hbar Ht)P_a \exp(-i\hbar H_a t) \tag{3.30a}$$

and similarly

$$\Omega_a^{(-)} = 1 + i \int_{-\infty}^{\infty} dt \; G^{(-)}(-t)H_a'G_a^{(+)}(t)P_a$$

$$= \lim_{t \to \infty} \exp(i\hbar Ht)P_a \exp(-i\hbar H_a t) \quad . \tag{3.30b}$$

These operators can be equivalently expressed in another integral form similar to (3.16 or 3.18). Without going through the detail, we present them below:

$$\Omega_a^{(+)} = \lim_{\varepsilon \to +0} \varepsilon \int_{-\infty}^{0} dt \; e^{\varepsilon t} \exp(i\hbar^{-1}Ht)P_a \exp(-i\hbar^{-1}H_a t) \tag{3.30c}$$

$$\Omega_a^{(-)} = \lim_{\varepsilon \to +0} \varepsilon \int_{0}^{\infty} dt \; e^{-\varepsilon t} \exp(i\hbar^{-1}Ht)P_a \exp(-i\hbar^{-1}H_a t) \quad . \tag{3.30d}$$

The S matrix is now defined as the operator that transforms the "in" state to the "out" state,

$$\psi_{b\beta}(t) = S_{ba}\psi_{a\alpha}(t) \quad . \tag{3.31}$$

It then follows from (3.29) and the definitions of the Møller wave operators that

$$S_{ab} = \Omega_b^{(-)\dagger}\Omega_a^{(+)} \tag{3.32a}$$

$$= \lim_{t \to -\infty, t' \to \infty} \exp(i\hbar^{-1}H_b t')P_b \exp[-i\hbar^{-1}H(t' - t)]P_a \exp(-i\hbar^{-1}H_a t) \quad . \tag{3.32b}$$

24

This S matrix differs from that for nonreactive scattering by the presence of the projection operators necessary for the limit to exist. It easily follows from (3.32b) that the S matrix so defined is unitary in an extended sense:

$$\sum_b S_{ab} S_{cb}^\dagger = P_a \delta_{ac} \qquad \text{and}$$

$$\sum_b S_{ba}^\dagger S_{bc} = P_a \delta_{ac} \quad .$$

These relations generalize the unitary relation of the S matrix for nonrearrangement scattering. The S matrix (3.32b) forms the starting point of a semiclassical theory of rearrangement collisions in Chap. 6.

3.2 Time-Independent Scattering Theory

For simplicity we shall review the theory in terms of diatomics (for more complete reviews, see [3.4,14,15]). The theory remains basically the same except for some minor modifications required for other types of molecules.

Imagine a diatomic molecule in collision with another in a force field. The internal states of the molecules in collision are denoted collectively by n (= $n_1,j_1,m_1;n_2,j_2,m_2$) where n_i denotes the vibrational quantum number, j_i is the rotational quantum number, and m_i is the z component of j_i of molecule i. The corresponding internal energy is expressed by E_n and the relative kinetic energy by E. The total energy of the system is then given by

$$\mathcal{E}_n = E_n + E \quad .$$

The orbital angular momentum of the relative motion is denoted by ℓ and its z component by m_ℓ. The wave function corresponding to the total angular momentum of the whole system J, its z component M, and the channel n satisfies the Schrödinger equation

$$H\psi_{n\ell}^{JM}(\mathbf{R},\mathbf{r}) = \mathcal{E}_n \psi_{n\ell}^{JM}(\mathbf{R},\mathbf{r}) \quad , \tag{3.33}$$

where the Hamiltonian consists of the relative kinetic energy H_0, the internal energy H_{int}, and the interaction energy $V(\mathbf{R},\mathbf{r})$,

$$H = H_0 + H_{int} + V(\mathbf{R},\mathbf{r})$$

$$H_0 = -(\hbar^2/2\mu)\nabla_R^2 \quad , \tag{3.34}$$

μ being the reduced mass of the relative motion, **R**, the relative distance between the colliding molecules, while **r** denotes collectively the internal coordinates $\mathbf{r}_1$, $\mathbf{r}_2$,... of atoms in the molecules in the body-fixed coordinate system. We assume that the Schrödinger equation for the internal motion,

$$H_{int}\phi_n(\mathbf{r}) = E_n\phi_n(\mathbf{r}) \quad , \tag{3.35}$$

is already solved. We also assume that the rearrangement channels are closed so that only elastic and inelastic scatterings are possible. It is, however, possible to formulate the theory with rearrangement channels open [3.3,4,16].

In the close-coupling theory we are formulating, the wave function is expanded into a set of internal wave functions,

$$\psi_{n\ell}^{JM} = \sum_{n'j'\ell'} R^{-1}u_{n'j'\ell'}^{nj\ell}(R)\psi_{n'j'\ell'}^{JM}(\hat{\mathbf{R}},\mathbf{r}) \quad . \tag{3.36}$$

Here $\hat{R}$ denotes the angular coordinates of the vector **R** and

$$\psi_{nj\ell}^{JM}(\hat{\mathbf{R}},\mathbf{r}) = \sum_{m_\ell=-\ell}^{\ell} \sum_{m=-j}^{j} C(j\ell mm_\ell|j\ell JM)Y_\ell^{m_\ell}(\hat{\mathbf{R}})\phi_n^{jm}(\mathbf{r}) \quad , \tag{3.37}$$

where $C(j\ell mm_\ell|j\ell JM)$ is the Clebsch-Gordan coefficient [3.17-21], $Y_\ell^m(\hat{\mathbf{R}})$ is the spherical harmonic, and $\phi_n^{jm}(\mathbf{r})$ is the internal wave function for the total rotational angular momentum state $j = j_1 + j_2$ and its z component m. We use here the *Condon-Shortley* notation [3.19] for the Clebsch-Gordan coefficient.

It must be noted that the internal wave function ϕ_n^{jm} may be written in terms of the radial wave functions and the angular wave function which may be written as a linear combination of spherical harmonics for rotational motions of molecules,

$$Y_j^m(\hat{\mathbf{r}}_1,\hat{\mathbf{r}}_2) = \sum_{m_1=-j_1}^{j_1} \sum_{m_2=-j_2}^{j_2} C(j_1j_2m_1m_2|j_1j_2jm)Y_{j_1}^{m_1}(\hat{\mathbf{r}}_1)Y_{j_2}^{m_2}(\hat{\mathbf{r}}_2) \quad .$$

Substitution of (3.36) into (3.33) and integration over the internal coordinates after multiplying the equation with $\psi_{nj\ell}^{JM*}$ yields a set of coupled differential equations for the relative radial wave functions,

$$\left[\frac{d^2}{dR^2} + k_{nn'}^2 - \ell'(\ell' + 1)/R^2 - \lambda^2 V_{n'j'\ell';n'j'\ell'}\right]u_{n'j'\ell'}^{nj\ell}(R)$$

$$= \lambda^2 \sum_{n''j''\ell''}' V_{n'j'\ell';n''j''\ell''}u_{n''j''\ell''}^{nj\ell}(R) \quad , \tag{3.38}$$

where

$$k^2_{nn'} = \lambda^2 (E_n - E_{n'})$$

$$V_{nj\ell;n'j'\ell'} = \int d\mathbf{R} \; d\mathbf{r} \; \phi_n^{jm*}(\mathbf{r}) Y_\ell^{m*}(\hat{R}) V(\mathbf{R},\mathbf{r}) \phi_{n'}^{j'm'}(\mathbf{r}) Y_\ell^{m'}(\hat{R}) \tag{3.39}$$

and the prime on the summation sign means the exclusion of the diagonal component from the sum. It is common that only open channels are taken into consideration in the theory. However, at this point there is no necessity for such a restriction. Solution of the coupled differential equations (3.38) furnishes the scattering matrix and all the information on the system considered. There are numerous papers on quantum-mechanical solution methods for (3.38) published in the past decades ([3.4] and references cited therein). In this work we shall study asymptotic solution methods along the line of the WKB method in quantum mechanics. Here, we assume that the solution of (3.38) is accomplished subject to the scattering boundary conditions on the radial wave functions $u_{n'j'\ell'}^{nj\ell}(R)$.

Then, the scattering matrix $S^J(nj\ell/n'j'\ell')$ may be defined by the requirement that the asymptotic form of the radial wave function be

$$u_{n'j'\ell'}^{nj\ell}(R) \sim \delta_{nn'}\delta_{jj'}\delta_{\ell\ell'} v_{nj}^{-1/2} \exp[-i(k_{nn'}R - \ell'\pi/2)]$$

$$-v_{n'j'}^{-1/2} S^J(nj\ell|n'j'\ell') \exp[i(k_{nn'}R - \ell'\pi/2)] \quad , \tag{3.40}$$

where v_{nj} is the velocity of the relative motion in channel (n,j). The scattering matrix is independent of M due to the isotropy of the space.

It is useful to define the following abbreviations for wave functions:

$$I_{n'j'nj\ell mm_\ell} = R^{-1} v_{nj}^{-1/2} \exp(-ik_{n'n}R) Y_\ell^m(\hat{R}) \phi_n^{jm}(\mathbf{r})$$

$$E_{n'j'nj\ell mm_\ell} = R^{-1} v_{nj}^{-1/2} \exp(ik_{n'n}R) Y_\ell^m(\hat{R}) \phi_n^{jm}(\mathbf{r}) \quad . \tag{3.41}$$

Then substitution of (3.40) into (3.36) yields the asymptotic formula for $\psi_{nj\ell}^{JM}$ in the form

$$\psi_{nj\ell}^{JM} \sim \sum_m \sum_{m_\ell} i^\ell C(j\ell mm_\ell | j\ell JM) I_{njnj\ell mm_\ell}$$

$$- \sum_{n'j'\ell'} \sum_{m'} \sum_{m'_\ell} i^{\ell'} S^J(nj\ell|n'j'\ell') C(j'\ell'm'm'_\ell | j'\ell'JM) E_{njn'j'\ell'm'm'_\ell} \quad . \tag{3.42}$$

The plane wave representing particles in state (nj) moving in the z direction may be asymptotically written as

$$v_{nj}^{-1/2} \exp(ik_{nn}z)\phi_n^{jm}(\mathbf{r}) \sim i\pi^{1/2}k_{nn'}^{-1} \sum_{j=0}^{\infty} \sum_{M=-J}^{J} \sum_{\ell=-|J-j|}^{J+j}$$

$$\sum_{m'=-j}^{j} \sum_{m'_\ell=-\ell}^{\ell} (2\ell + 1)^{1/2} i^\ell C(j\ell m0|j\ell JM)C(j\ell m'm'_\ell|j\ell JM)$$

$$\times (i^\ell I_{njnj\ell mm_\ell} - i^{-\ell}E_{njnj\ell mm_\ell}) \quad , \tag{3.43}$$

where we used the expansion theorem for the plane wave, the orthogonality
relation of the Clebsch-Gordan coefficients, and (3.37). The wave function
describing the collision subject to the scattering boundary condition is then
obtained by taking a linear combination of (3.41) and making use of (3.42):

$$\Psi_{njm}(\mathbf{R},\mathbf{r}) \sim i\pi^{1/2}k_{nn}^{-1} \sum_{J=0}^{\infty} \sum_{M=-J}^{J} \sum_{\ell=|J-j|}^{J+j} (2\ell + 1)^{1/2} i^\ell C(j\ell m0|j\ell JM)\Psi_{nj\ell}^{JM}$$

$$\sim i\pi^{1/2}k_{nn}^{-1} \sum_J \sum_M \sum_\ell \sum_{m'} \sum_{m_\ell} (2\ell + 1)^{1/2} i^\ell C(j\ell m0|j\ell JM)C(j\ell m'm'_\ell|j\ell JM)$$

$$\times (i^\ell I_{njnj\ell mm_\ell} - i^{-\ell}E_{njnj\ell mm_\ell})$$

$$+ \sum_{n'j'\ell'} i^{-\ell'}[\delta_{nn'}\delta_{jj'}\delta_{\ell\ell'} - S^J(nj\ell|n'j'\ell')]E_{njn'j'\ell'm'm'_\ell} \quad .$$

$$\tag{3.44}$$

The second part of (3.44) represents the scattered waves in various channels, elastic and inelastic. By rearranging terms in the above expression,
we cast (3.44) in the form

$$\Psi_{njm} \sim v_{nj}^{-1/2} \exp(ik_{nn}z)\phi_n^{jm}(\mathbf{r})$$

$$+ i\pi^{1/2}k_{nn}^{-1} \sum_{n'j'm'} U(njm|n'j'm')v_{n'j'}^{-1/2}R^{-1} \exp(ik_{nn'}R)\phi_{n'}^{j'm'}(\mathbf{r}) \tag{3.45}$$

from which we identify the scattering amplitudes for various processes:

$$U(njm|n'j'm') = \sum_{j=0}^{\infty} \sum_{M=-J}^{J} \sum_{\ell=-|J-j|}^{J+j} \sum_{\ell'=-|J-j'|}^{J+j'} \sum_{m'_\ell=-\ell'}^{\ell'} (2\ell + 1)^{1/2} i^{\ell-\ell'}$$

$$\times C(j\ell m0|j\ell JM)[\delta_{nn'}\delta_{jj'}\delta_{\ell\ell'} - S^J(nj\ell|n'j'\ell')]$$

$$\times C(j'\ell'm'0|j'\ell'JM)Y_{\ell'}^{m'_\ell}(\hat{\mathbf{R}}) \quad . \tag{3.46}$$

The amplitude of the wave in the channel $(n'j'\ell')$ scattered into solid angle $R^2 v_{n'j'} d\Omega$ in the unit time defines the differential cross section [3.22]

$$d\sigma(njm|n'j'm')/d\Omega = \pi k_{nn'}^{-2}|U(njm|n'j'm')|^2 \quad . \tag{3.47}$$

Experiment is usually done such that the angular-momentum polarizations are not measured. In that case, the observed differential cross section is the average of (3.47) over the initial polarizations m which is then summed over all the final polarizations m':

$$d\sigma(nj|n'j')/d\Omega = (2j + 1)^{-1} \sum_m \sum_{m'} d\sigma(njm|n'j'm')/d\Omega \quad . \tag{3.48}$$

By substituting (3.47,46) in turn, we may express the differential cross section as

$$\frac{d\sigma(nj|n'j')}{d\Omega} = \frac{\pi}{(2j + 1)k_{nn'}^2}\left|\sum_{\ell\ell'm_\ell'} (2\ell + 1)^{1/2} i^{\ell-\ell'} B(\ell\ell'm_\ell') Y_\ell^{m_\ell}(\hat{R})\right|^2 \tag{3.49}$$

where

$$B(\ell\ell'm_\ell') = \sum_J \sum_M C(j\ell m0|j\ell JM)[\delta_{nn'}\delta_{jj'}\delta_{\ell\ell'} - S^J(nj\ell|n'j'\ell')]$$

$$\times C(j'\ell'm'm_\ell'|j'\ell'JM) \quad . \tag{3.50}$$

These formulas for differential cross sections may be cast into somewhat different forms by following *Blatt* and *Biedenharn* [3.22], but since the forms are completely equivalent, they are not presented here, and the reader is referred to [3.22].

What is important for us here is the fact that the observable cross sections are given in terms of the scattering matrix elements $S^J(nj\ell|n'j'\ell')$ and their determination is the major task of scattering theory and of this monograph. The way the theory is formulated in this section, the scattering matrix is completely determined from the solutions of (3.38) subject to the scattering boundary conditions. Equation (3.38) represents, as it stands, an infinite set of coupled equations. Unless the energy is infinite, most of the states are not energetically accessible. Such states are called closed channels; the rest, open channels. Since the translational energies are negative for closed channels, the corresponding acceptable wave functions are exponentially decreasing functions characteristic of bound states. Since bound states are localized in space, such states do not directly contribute to the scattering amplitudes, although they influence the collision indirectly.

The common practice in scattering theory is to truncate the complete set of wave functions to include only the open channels or only a few closed channels in addition to the open channels, thereby ensuring the desired accuracy. The latter approach is often taken in numerical solution methods [3.23-26] in which the number of closed channels included is usually determined empirically by inspecting the rate of convergence of the solutions as the number of close channels included is increased. In the subsequent chapters most analyses have been done with open channels only.

4. Elastic Scattering

Since it provides rich information on atoms and molecules interacting at relatively low energies and is responsible for many thermodynamic and transport phenomena, elastic scattering is not only of practical interest in its own right, but also has provided exploratory ground for theorists aiming to develop theories for more complicated scattering phenomena. The theories now have been well worked out in many directions [4.1-19], but we shall take up the subject here to develop some basic theoretical tools to be applied later to some problems of practical interest. We shall first consider the simple WKB approximation to elastic scattering and then discuss corrections of the WKB result by using uniform semiclassical wave functions, which simplify discussions of multiturning-point problems. Since one of the aims of this monograph is to develop approximation formulas useful for experimentalists, we shall also consider some procedures for calculating phase shifts, given a potential function. The differential equations of interest here are either a one-dimensional Schrödinger equation in the Cartesian coordinate representation or the radial Schrödinger equation in the spherical coordinate representation. We consider the WKB approximation for the Schrödinger equation in the prolate-spheroidal-coordinate system in Chap. 11, where scattering by an ellipsoid particle is discussed.

4.1 The WKB Approximation for Phase Shifts

Let us assume that two particles of reduced mass m undergo a collision under the influence of a spherically symmetric interaction potential $V(r)$ which is a finite-ranged function of distance r. Then the equation of motion of interest is the Schrödinger equation for the radial wave function $u_\ell(r)$,

$$\left\{ \frac{d^2}{dr^2} + \lambda^2 \left[E - \frac{\ell(\ell + 1)}{\lambda^2 r^2} - V(r) \right] \right\} u_\ell(r) = 0 \quad , \tag{4.1}$$

where

$$\lambda = (2m)^{1/2}/\hbar$$

and ℓ is the orbital angular momentum quantum number. The wave number of the relative motion will be denoted by $k = \lambda E^{1/2}$.

Since the collision process has cylindrical symmetry, say, around the z axis of the coordinate system that is parallel to the direction of the incident plane wave, the full wave function $\Psi(\mathbf{r})$ does not depend on the z component of the orbital angular momentum and therefore is given by

$$\Psi(\mathbf{r}) = \sum_{\ell=0}^{\infty} r^{-1} u_\ell(r) P_\ell(\cos\theta) \quad . \tag{4.2}$$

The radial wave functions are subject to the scattering boundary conditions

$$u_\ell(r) \sim (\pi k)^{-1/2} \sin(kr - \ell\pi/2 + n_\ell) \quad , \tag{4.3}$$

where n_ℓ is the phase shift for the ℓth partial wave.

Then, following the procedure described in Sect.3.2, we find the scattering amplitude for elastic scattering,

$$f_k(\theta) = (2ik)^{-1} \sum_{\ell=0}^{\infty} (2\ell + 1)(S_\ell - 1) P_\ell(\cos\theta) \quad , \tag{4.4}$$

where the scattering matrix element is now given as

$$S_\ell = \exp(2in_\ell) \quad . \tag{4.5}$$

The scattering amplitude is therefore fully known if the scattering phase shifts n_ℓ have been calculated for all ℓ. They can be found if (4.1) is solved subject to (4.3). Here we calculate them by solving (4.1) by the WKB method.

In the WKB approximation method the centrifugal term in (4.1) is replaced by

$$(\ell + 1/2)^2/\lambda^2 r^2 \quad ,$$

which is called the Langer modification [2.18].

If the potential behaves like r^{-n}, $n < 2$, near $r = 0$, then $u_\ell \sim r^{\ell+1}$ near the origin. But with the centrifugal term unmodified the regular WKB solution is $u_\ell \sim r^{[\ell(\ell+1)]^{1/2}+1/2}$. This is corrected if the Langer modification is made, which yields $u_\ell \sim r^{\ell+1}$. Therefore the reason for such a modification is purely mathematical, enabling the WKB approximation to give a correctly behaving solution at $r = 0$. This is necessary if the potential is not singular, but unnecessary if the potential is singular, i.e.,

$$V(r) \sim r^{-n}, \quad n > 2 \quad,$$

since then the behavior of the wave function at $r = 0$ is determined by the potential, not by the centrifugal term that has a weaker singularity at the origin than the potential. Since the potentials relevant to molecular collisions are mostly singular, e.g., the Lennard-Jones potentials, the modification is not essential. Nevertheless, we introduce it for all types of potentials for a uniform presentation, especially in view of the fact that it makes only small differences to the phase shifts.

In (4.1) λ plays the role of the large parameter appearing in (2.5). Therefore, we may expand the solution in an asymptotic series in λ. By applying the procedure leading to (2.16), we thus obtain the general WKB solution for u_ℓ,

$$u_\ell(r) = p^{-1/2}(r)[C \exp(-i\xi) + D \exp(i\xi)] \tag{4.6}$$

where C and D are constants,

$$p(r) = [E - V(r) - (\ell + 1/2)^2/\lambda^2 r^2]^{1/2} \tag{4.7}$$

and

$$\xi = \lambda \int_{r_0}^{r} dr' \, p(r') \quad. \tag{4.8}$$

Here r_0 is the turning point defined by $p^2(r_0) = 0$.

In (4.6) we have retained only the leading approximation in (2.15), i.e., $S_0^{(\pm)}$. Here we specifically consider only the case of a single turning point, that is, when $p^2(r)$ has only one real positive zero r_0. Then the region $r < r_0$ is classically forbidden, while the region of $r > r_0$ is classically accessible. Obviously, (4.6) is not uniformly valid for all values of r and we have a connection problem in hand. Since the local behavior of the solution in the neighborhood of r_0 is of interest here, this problem can be reduced to that of the asymptotic expansions for the Airy functions discussed in detail in Sect.2.3. First, observe that $p^2(r)$ can be expanded around r_0, and in the neighborhood sufficiently close to r_0 only the linear term may be retained:

$$\lambda^2 p^2(r) = a(r - r_0) \equiv -z \quad (a > 0) \quad. \tag{4.9}$$

Consequently, in the neighborhood of r_0 the integral in (4.8) is easily integrated to

$$\xi = i2z^{3/2}/3$$

and the wave function may be written as

$$u_\ell = C(0,z) + D(z,0) \quad , \tag{4.6a}$$

where we are using the notations $(0,z)$ and $(z,0)$ defined in Chap. 2 for wave functions. Note that the definition of z by (4.9) makes the positive real axis in the z plane the classically forbidden region. The z plane is cut just under the negative real axis (Fig.4.1). The situation now is exactly the same as for (2.32). Therefore, the connection formulas established by means of Airy functions may be applied and the tracing rules in Sect.2.4 become relevant.

Since the wave function must be equal to zero at $r = 0$ and the acceptable solution is the second term in (4.6a), C must be set equal to zero so that

$$u_\ell = D(z,0) \tag{4.10}$$

in the classically forbidden region, i.e., $\mathrm{Re}\{z\} > 0$. The solution is subdominant there. When this solution is traced around to the negative real axis in the z plane, on the ray $\theta = 0$ we obtain

$$u_\ell = D2^{-1}\lambda^{1/2}[\exp(i\pi/4)(0,\rho) + \exp(-i\pi/4)(\rho,0)] \quad , \tag{4.11}$$

where ρ is the amplitude of z,

$$\left.\begin{array}{l} z = \rho \, \exp(i\theta) \\[2mm] (0,\rho) = \rho^{-1/2}\exp(-i2\rho^{3/2}/3) \\[2mm] (\rho,0) = \rho^{-1/2}\exp(i2\rho^{3/2}/3) \end{array}\right\} \quad . \tag{4.12}$$

Now reverting back to the original variables and with (4.8) we finally obtain the wave function in the classically accessible region

$$u_\ell = Dp^{-1/2}(r) \sin\left[\lambda \int_{r_0}^{r} dr' p(r') + \pi/4\right] \quad . \tag{4.13}$$

Taking the asymptotic form of this solution at large distance and comparing it with (4.3), we identify the phase shift in the WKB approximation

$$n_\ell^{WKB} = 2^{-1}\pi(\ell + 1/2) - kr_0 + \int_{r_0}^{\infty} dr[\lambda p(r) - k] \quad . \tag{4.14}$$

This formula may be used in (4.4) to calculate the WKB scattering amplitude if the energy is sufficiently high.

For high energies (typically $E \gtrsim |V_{min.}|$, the well depth) the high-ℓ phase shifts are sufficiently accurate (ca. 0.1%) to satisfy the need of most experimental data [4.20,21]. High-resolution experiments, however,

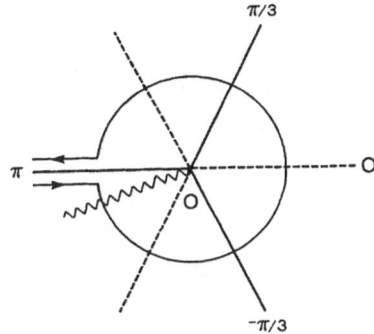

<u>Fig.4.1.</u> Tracing path. The solid lines are the anti-Stokes lines, the dotted, the Stokes lines. The wavy line is the branch cut

might find the WKB phase shifts a little wanting in accuracy. In any event, (4.14) does not hold for the entire energy range, but breaks down if more than one turning point arises as the energy gets lower than a certain value, e.g., the centrifugal barrier in the effective potential. Such a situation requires modification of the asymptotic solution method. Before we consider this subject and corrections to (4.14), we indicate a computational proce- dure for the WKB phase shifts.

The phase integral in (4.14) is not possible to calculate analytically except for a few special types of potentials which are usually not interest- ing physically. Consequently, a numerical method is required. The computa- tion is greatly facilitated if the range of integration is divided into two, say, at $r = r_m$, where r_m is chosen such that the potential is practically equal to zero beyond the point, and then if the integration for $r \geq r_m$ is performed analytically, while the integration for $r < r_m$ is done numeri- cally. To discuss this point, it is convenient to recast (4.14) in the form

$$n_\ell^{WKB} = \lambda \int_{r_0}^\infty dr\, p(r) - \lambda \int_{r_m}^\infty dr\, p_0(r)$$

$$= \lambda \int_{r_0}^{r_m} dr\, p(r) + \lambda \int_{r_m}^\infty dr[p(r) - p_0(r)] \quad,$$

where $p_0 = [E - (\ell + 1/2)^2/\lambda^2 r^2]^{1/2}$, and r_m is the root of $p_0^2(r)$. The first integral in the second line above is evaluated numerically and the second integral analytically by the method [4.21] originally used by *Massey* and *Mohr* [4.22]. If $p(r)$ is expanded in series of $V(r)$, then to first order in $V(r)$ the integral becomes

$$\Delta n_\ell^{WKB} \equiv \lambda \int_{r_m}^\infty dr[p(r) - p_0(r)]$$

$$= -(\lambda/2E^{1/2}) \int_{r_m}^{\infty} dr\ V(r)[1 - (\ell+1/2)^2/\lambda^2 E r^2]^{-1/2} \ .$$

It is generally possible to evaluate the integral analytically. For example, if $V(r) = -c^{(n)}/r^n$ at large r, then it is shown [4.20,22] that

$$\Delta\eta_\ell^{WKB} = \frac{\pi^{1/2}\lambda^2 c^{(n)} k^{n-2} \Gamma(n/2 - 1/2)}{4(\ell + 1/2)^{n-1}\Gamma(n/2)} \ .$$

This Massey-Mohr formula has proven useful for computing phase shifts.

4.2 Corrections to the WKB Phase Shifts

Corrections to the first-order WKB solution have been reported in [4.2,6-8, 12,18,23]. All of them compute terms of $O(\lambda^{-1})$ in the asymptotic expansion in one way or another. In the present formulation, corrections to the WKB phase shifts can be made if the higher-order members of the hierarchy in (2.14) are solved. Thus the second-order correction will be obtained if $S_2^{(\pm)}$ are calculated, which may be written in the notations used here as

$$S_2^{(\pm)} = \mp (i/2) \int_0^\xi d\xi' \{p^2(\xi'),r\} \ . \tag{4.15}$$

Since in the vicinity of the turning point, i.e., $\xi = 0$,

$$\{p^2(\xi),r\} \sim \xi^{-2} \ ,$$

the integral diverges—and this is a general feature if the hierarchy (2.14) is integrated successively. A trick used to circumvent it is an analytic continuation into the complex plane [4.6] and to distort the integral into a contour integral in the cut complex plane. The result obtained is now well defined, but requires complex integration. However, it is possible to avoid such a procedure [4.2,5,12] and we shall discuss it here, using the method employed in [4.12]. The corrections we obtain, however, are not exactly second order, but include higher orders as well.

Let us observe that the turning-point difficulty can be circumvented if uniform WKB solutions [2.6,4.24,25] are used. They can be introduced into the theory as follows. The radial Schrödinger equation (4.1) is recast in the form

$$(d^2/dr^2 + \lambda^2\phi\phi'^2 + 2^{-1}\{\phi,r\})u_\ell(r) = 2^{-1}\{\phi,r\}u_\ell(r) \ , \tag{4.16}$$

where $\{\phi,r\}$ is a Schwartzian derivative,

$$\{\phi,r\} = (\phi'''/\phi') - (3/2)(\phi''/\phi')^2 \ , \quad \phi' = (d\phi/dr) \ , \quad \text{etc.} \ , \qquad (4.17)$$

and the new variable ϕ is chosen such that

$$\phi\phi'^2 = p^2(r) \ . \qquad (4.18)$$

That is, ϕ is determined by the integrals

$$(2/3)\phi^{3/2} = \int_{r_0}^{r} dr'p(r') \ , \quad (r > r_0) \qquad (4.19a)$$

$$(2/3)(-\phi)^{3/2} = \int_{r}^{r_0} dr'|p(r')| \ , \quad (r < r_0) \ . \qquad (4.19b)$$

The significance of writing the differential equation in the form (4.16) can be understood if we consider the homogeneous part of (4.16);

$$(d^2/dr^2 + \lambda^2\phi\phi'^2 + 2^{-1}\{\phi,r\})Y(r) = 0 \ . \qquad (4.20)$$

Here, if we set

$$Y(r) = (\phi')^{-1/2}W(\phi) \ , \qquad (4.21)$$

$$t = \lambda^{2/3}\phi(r) \ , \qquad (4.22)$$

then it is easy to see that (4.20) transforms into the differential equation

$$(d^2/dt^2 + t)W(t) = 0 \qquad (4.23)$$

which is the well-known differential equation for the Airy functions. The general solution to (4.20) is then

$$W(t) = C_1 Ai(-t) + C_2 Bi(-t) \ .$$

Therefore, if we ignore the right-hand side of (4.16) which is $O(\lambda^0)$, we find the approximate solution to (4.1) satisfying the boundary condition at $r = 0$ is

$$u_\ell(r) = C(t')^{-1/2}Ai(-t) \qquad (4.24)$$

$$\sim C(\pi k)^{-1/2} \sin\left[\lambda \int_{r_0}^{r} dr'p(r') + \pi/4\right] \ . \qquad (4.25)$$

When compared with the scattering boundary condition (4.3), this asymptotic formula yields the WKB phase shift n_ℓ^{WKB}. In view of this result we now see that the correction to the WKB phase shift will be obtained if (4.16) is solved by treating the Airy functions as its homogeneous solutions. We approach the problem in the following manner.

First define

$$A(r) = (t')^{-1/2} Ai(-t) \qquad (4.26a)$$

$$B(r) = (t')^{-1/2} Bi(-t) \quad , \qquad (4.26b)$$

and then write the wave function in terms of $A(r)$ and $B(r)$:

$$u_\ell(r) = \alpha_\ell(r) A(r) + \beta_\ell(r) B(r) \quad , \qquad (4.27)$$

where $\alpha_\ell(r)$ and $\beta_\ell(r)$ are so far unknown functions that must be determined. To obtain the differential equations for them, we differentiate (4.27) under the condition

$$A \frac{d\alpha_\ell}{dr} + B \frac{d\beta_\ell}{dr} = 0 \qquad (4.28)$$

so that we obtain

$$\frac{du_\ell}{dr} = \alpha_\ell \frac{dA}{dr} + \beta_\ell \frac{dB}{dr} \quad . \qquad (4.29)$$

The condition (4.28) is similar to that used by *Kemble* [2.17] and *N. Fröman* and *P.O. Fröman* [4.9] in their discussions on the WKB connection formulas. The condition is imposed to make α_ℓ and β_ℓ change as slowly as possible. When (4.29) is differentiated once more and when (4.16,18,20) are used, the following differential equation is obtained:

$$\frac{dA}{dr} \frac{d\alpha_\ell}{dr} + \frac{dB}{dr} \frac{d\beta_\ell}{dr} = 2^{-1}\{\phi,r\}(\alpha_\ell A + \beta_\ell B) \quad . \qquad (4.30)$$

This equation together with (4.28) form a set of two equations for the un-determined functions α_ℓ and β_ℓ. Since the Wronskian of the Airy functions is

$$Ai(-t) \frac{d}{dt} Bi(-t) - Bi(-t) \frac{d}{dt} Ai(-t) = \frac{1}{\pi} \quad ,$$

we find

$$A(r) \frac{d}{dr} B(r) - B(r) \frac{d}{dr} A(r) = -\frac{1}{\pi} \quad .$$

When this identity is used in (4.28,30), the first-order coupled equations may be put into the following form:

$$\frac{d}{dr} \mathbf{X}(r) = \mathbf{W}(r)\mathbf{X}(r) \quad , \qquad (4.31)$$

where the column vector $\mathbf{X}$ and matrix $\mathbf{W}$ are respectively defined by the re-lation

$$\mathbf{X}(r) = \{\alpha_\ell(r) , \beta_\ell(r)\} \quad , \qquad (4.32)$$

$$W(r) = (\pi/2)\{\phi, r\} \begin{pmatrix} AB & B^2 \\ -A^2 & -AB \end{pmatrix} \; . \tag{4.33}$$

Since $u_\ell = 0$ at $r = 0$, the functions α_ℓ and β_ℓ must satisfy the boundary condition

$$\alpha_\ell(0) = \alpha_0 \; , \qquad \beta_\ell(0) = 0 \; ,$$

that is, at $r = 0$

$$X(r = 0) \equiv X_0 = \{\alpha, \; 0\} \; . \tag{4.34}$$

Solution of (4.31) provides not only the functions α_ℓ and β_ℓ, but also u_ℓ in the end. Moreover, the asymptotic values of α_ℓ and β_ℓ directly supply the correction to the WKB phase shift, as will be shown.

To derive the correction formula, we assume for the moment that (4.31) is solved and hence α_ℓ and β_ℓ are known. We then take the asymptotic form for $u_\ell(r)$ given by (4.27),

$$u_\ell(r) \sim (\pi k)^{-1/2} \{\alpha_\ell(\infty)\sin[\chi(r) + \pi/4]$$
$$+ \beta_\ell(\infty)\cos[\chi(r) + \pi/4]\} \tag{4.35}$$

where

$$\chi(r) = \lambda \int_{r_0}^{r} dr' p(r') \; .$$

When (4.35) is compared with (4.3), the phase shift n_ℓ is obtained,

$$n_\ell = n_\ell^{WKB} + \Delta_\ell \; , \tag{4.36}$$

where the correction to the WKB phase shift is now given by

$$\Delta_\ell = \tan^{-1}[\beta_\ell(\infty)/\alpha_\ell(\infty)] \; . \tag{4.37}$$

The correction is entirely determined by the solution of the coupled equations (4.31).

To facilitate practical solution of (4.31), we express X in the exponential form[1]

---

1 It is also possible to write X in a product integral

$$X(r) = \prod_{r_0}^{r} \exp[W(s)ds]X(r_0)$$

which is also a formal solution to (4.31) and yields the first Magnus approximation to be shown, if $[W(r_i), W(r_j)] = 0$ for $i \neq j$. For product integrals see [4.26-28]. One useful fact of the theory of product integration is the strong analogy between this theory and the usual theory of Riemann integration. Sums in ordinary integration go over to products in the product integration theory. Likewise, the additive neutral element 0 goes over to the multiplicative neutral element I and the additive inverse $-B$ goes over to the multiplicative inverse B^{-1}.

$$\mathbf{X} = \exp(\mathbf{F})\mathbf{X}_0 \quad . \tag{4.38}$$

The square matrix $\mathbf{F}$ can be shown [4.29-32] to satisfy the differential equation

$$\frac{d\mathbf{F}}{dr} = \{\mathbf{F}(\exp \mathbf{F} - 1)^{-1}, \mathbf{W}\}$$

$$= \mathbf{W} - 2^{-1}[\mathbf{F},\mathbf{W}] + \sum_{j=1}^{\infty} \frac{(-)^{j-1}B_j}{(2j)!} \{\mathbf{F}^{2j},\mathbf{W}\}$$

$$\equiv \mathbf{M}(\mathbf{F}) \quad , \tag{4.39}$$

where B_j is the jth Bernoulli number [2.2] and

$$[\mathbf{F},\mathbf{W}] = \mathbf{FW} - \mathbf{WF} \quad , \quad \{\mathbf{F}^2,\mathbf{W}\} = [\mathbf{F},[\mathbf{F},\mathbf{W}]] \, , \quad \text{etc.}$$

Equation (4.39) is a nonlinear transformation solvable by iteration. That is, we define a sequence $\{\mathbf{F}_0 = 1, \mathbf{F}_1, \mathbf{F}_2, \ldots\}$ such that

$$\frac{d\mathbf{F}_m}{dr} = \mathbf{M}(\mathbf{F}_{m-1}) \quad .$$

With the condition $\mathbf{F}_m(r_0) = 0$ this equation is integrated to

$$\mathbf{F}_m = \int_{r_0}^{r} dr \, \mathbf{M}[\mathbf{F}_{m-1}(r)] \quad . \tag{4.40}$$

The existence of the limit of the sequence is assumed

$$\mathbf{F}(r) = \lim_{m \to \infty} \mathbf{F}_m(r) \quad .$$

Thus the sequence of solution is

$$\mathbf{X}^{(m)} = \exp(\mathbf{F}_m)\mathbf{X}_0 \quad . \tag{4.41}$$

This is the mth approximant to $\mathbf{X}$ under the iteration scheme.

The calculation of the infinite series in (4.39) is formidable in general, but we are lucky in the present case, since it is possible to sum the infinite series. The proof of summability is based on the fact that $\mathbf{F}_m$ is involutional[2], i.e.,

$$\mathbf{F}_m^2 = -\lambda_m \mathbf{I} \quad , \tag{4.42}$$

---

[2] A matrix $\mathbf{F}$ is called involutional if $\mathbf{F}^2 = f\mathbf{I}$ where f is a number. Well-known examples of involution matrices are the Pauli spin matrices and the time-reveral operator $\boldsymbol{\theta}$; $\theta^2 = \mathbf{I}$.

where I is the two-dimensional unit matrix and λ_m is given in terms of the elements of F_m. Since involution matrices must be traceless, F_m may be given the form

$$F_m = \begin{pmatrix} f_m & g_m \\ h_m & -f_m \end{pmatrix} \quad . \tag{4.43}$$

Squaring this matrix, we find

$$\lambda_m = -(f_m^2 + g_m h_m) \quad . \tag{4.44}$$

The fact that F_m is involutional and has the form given by (4.43) is proved in [4.12] to which the reader is referred; see also Sect.7.2.3. Since the involutionality of F_{m-1} implies the identity

$$\{F_{m-1}^{2n}, W\} = (-4\lambda_{m-1})^{n-1}\{F_{m-1}^2, W\} \quad ,$$

which may be proved by induction, we may sum the infinite series in (4.39) into a rather simple form

$$\sum_{j=1}^{\infty} \frac{(-)^{j-1}B_j}{(2j)!} \{F_{m-1}^{2j}, W\} = (4\lambda_{m-1})^{-1}[1 - \lambda_{m-1}^{1/2} \cot\lambda_{m-1}^{1/2}]\{F_{m-1}^2, W\} \tag{4.45}$$

for which the Bernoulli expansion [2.2] is used. Therefore, we now have for the mth iterate

$$F_m = F_1 - 2^{-1} \int_{r_0}^{r} dr[F_{m-1}, W]$$

$$+ \int_{r_0}^{r} dr(4\lambda_{m-1})^{-1}(1 - \lambda_{m-1}^{1/2} \cot\lambda_{m-1}^{1/2})\{F_{m-1}^2, W\} \quad . \tag{4.46}$$

When the right-hand side of this equation is explicitly worked out and compared with (4.43), the elements of F_m can be identified.

The exponential of the involution matrix can now be cast in the form

$$X^{(m)} = (\cos\lambda_m^{1/2}I + \lambda_m^{-1/2} \sin\lambda_m^{1/2}F_m)X_0 \quad . \tag{4.47}$$

This is the mth approximant for X that may be used for computation. If we take m = 1, then we obtain the first-order Magnus approximation

$$F_1 = \int_{r_0}^{r} dr' \, W(r')$$

and the corresponding correction to the WKB phase shift Δ_ℓ'

$$\Delta'_\ell = -\tan^{-1}\left\{\lambda_{1\infty}^{-1/2}(A^2)_\infty \tan\lambda_{1\infty}^{1/2}[1 + \lambda_{1\infty}^{1/2}(AB)_\infty \tan\lambda_{1\infty}^{1/2}]^{-1}\right\} \quad , \tag{4.48}$$

where

$$(A^2)_\infty = (\pi/2) \int_{r_0}^{\infty} dr\{\phi,r\}A^2(r) \quad , \quad \text{etc.} \quad , \tag{4.49}$$

$$\lambda_{1\infty} \equiv \lambda_1(\infty) = (A^2)_\infty(B^2)_\infty - (AB)_\infty^2 \quad . \tag{4.50}$$

The stationary-phase method for integrals shows that

$$|(AB)_\infty| \ll (A^2)_\infty$$

and

$$(A^2)_\infty \approx (B^2)_\infty \quad .$$

Therefore, if $(AB)_\infty$ is neglected in comparison with $(A^2)_\infty$ and $(B^2)_\infty$, we obtain

$$\lambda_{1\infty}^{1/2} \approx (A^2)_\infty$$

and

$$\eta_\ell = \eta_\ell^{WKB} + \Delta'_\ell \quad , \tag{4.51}$$

$$\Delta'_\ell \approx -(A^2)_\infty = -(\pi/2)\int_{r_0}^{\infty} dr\{\phi,r\}A^2(r) \quad . \tag{4.52}$$

This integral is well behaved and free of divergence thanks to the presence of the function $A(r)$ and the new variable ϕ, despite the fact that it also involves a Schwartzian derivative as does (4.15). It is fairly easy to compute numerically. To show its accuracy, in Table 4.1 we compare phase shifts for the Lennard-Jones (12,6) potentials with numerical solution phase shifts.

In the Table η' are the phase shifts calculated with (4.51) and η'' are the phase shifts calculated with a correction from the classically forbidden region added to (4.51). We have not considered the latter here (see [4.12] for it).

4.3 The WKB Phase Shifts at Low Energies

If the potential function has an attractive branch and ℓ is sufficiently large at low energies, then $p^2(r)$ can have more than one real root and there are possible quasi-bound states which are temporarily trapped in the effective potential well before they decay into the continuum. In the language of scattering theory we now have resonance states. The scattering process in-

Table 4.1. Phase shifts for potential $V = 4\varepsilon[(\sigma/r)^{12} - (\sigma/r)^6]$, $k\sigma = 10$, $2m\varepsilon\sigma^2/\hbar^2 = 125$. The last column contains the phase shifts calculated numerically by *Bernstein* [4.33] (quoted with permission from the author)

ℓ	η^{WKB}	η'	η''	η_{exact}
0	-6.684	-6.725	-6.725	-6.716
1	-5.208	-5.249	-5.249	-5.24
2	-3.827	-3.868	-3.867	-3.86
3	-2.542	-2.584	-2.583	-2.572
4	-1.354	-1.396	-1.395	-1.38
5	-0.266	-0.309	-0.309	-0.297
6	+0.720	+0.676	+0.676	+0.68
7	1.600	1.555	1.555	1.56
8	2.371	2.324	2.325	2.336
9	3.026	2.977	2.978	3.02
10	3.559	3.508	3.508	3.521
11	3.959	3.904	3.905	3.90
12	4.214	4.155	4.155	4.168
13	4.305	4.240	4.241	4.254
14	4.204	4.133	4.135	4.150
15	3.862	3.786	3.789	3.805
16	3.180	2.969	2.975	2.977
17	1.708	-	-	1.550
18	0.887	0.903	0.910	0.919
19	0.619	0.621	0.626	0.64
20	0.457	0.457	0.459	0.468
21	0.348	0.346	0.347	0.35
22	0.271	0.268	0.269	0.27
23	0.214	0.211	0.212	0.21
24	0.172	0.169	0.170	0.17
25	0.140	0.137	0.138	0.142

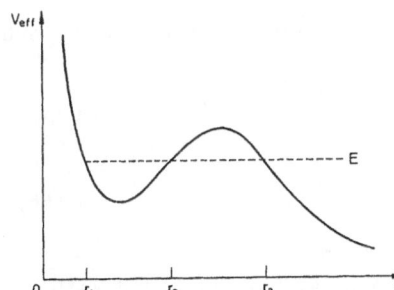

Fig.4.2. An effective potential. The classical turning points are denoted by r_1, r_2, and r_3

volving such states is called resonance scattering, and may be treated in the semiclassical-theory framework. It is possible to use the tracing rules presented in the previous chapter to deal with resonance scattering of this kind, but we shall use a uniform semiclassical approach similar to that in Sect.4.2, since the method yields not only the phase shifts including those

obtainable through the tracing method [2.13,4.9,34], but it also indicates
a way to improve the results presented below.

Imagine an effective potential as depicted schematically in Fig.4.2 and
assume that the relative kinetic energy is between the minimum and maximum
of V_{eff} where

$$V_{eff}(r) = V(r) + (\ell + 1/2)^2/\lambda^2 r^2 \quad .$$

Uniform semiclassical wave functions were introduced in the form of Airy
functions in Sect.4.2. They were reasonable since $p^2(r)$ had a single
turning point at the given range of energy. Since now there is more than
one turning point, such uniform semiclassical wave functions are no longer
very useful and consequently the radial Schrödinger equation must be cast
differently from (4.16). It can be done by following the idea of *Miller* and
Good [4.35] and others [4.24,25,36]. We thus recast (4.1) as follows:

$$[d^2/dr^2 + \lambda^2 f(\phi)\phi'^2 + 2^{-1}\{\phi,r\}]u_\ell(r) = 2^{-1}\{\phi,r\}u_\ell(r) \quad , \tag{4.53}$$

where $f(\phi)$ is an analytic function of ϕ such that

$$f(\phi)\phi'^2 = p^2(r) \quad , \tag{4.54}$$

with the precise form for $f(\phi)$ to be specified shortly. When that is done,
(4.54) completely determines the transformation from the variable r to the
new variable ϕ.

As in the case of a single turning point considered before, the basis of
the analysis is the homogeneous differential equation

$$[d^2/dr^2 + \lambda^2 f(\phi)\phi'^2 + 2^{-1}\{\phi,r\}]Y(r) = 0 \quad . \tag{4.55}$$

We now divide the whole range of the variable r into two parts $[0,r_2]$ and
$[r_2,\infty]$ and define $f(\phi)$ such that

$$f(\phi) = c^2 - \phi^2 \quad \text{for} \quad [0,r_2] \tag{4.56a}$$

and

$$f(\phi) = \phi^2 - c^2 \quad \text{for} \quad [r_2,\infty] \quad , \tag{4.56b}$$

where c is a constant whose meaning is given below.

We consider the region $[0,r_2]$ first. Equation (4.56a) means that with

$$t = (2\lambda)^{1/2}\phi \tag{4.57}$$

and

$$a = \lambda c/2 \tag{4.58}$$

the variable t is connected to r by

$$2^{-1} \int_{-2\sqrt{a}}^{t} ds(4a - s^2)^{1/2} = \lambda \int_{r_1}^{r} p(r')dr' \quad . \tag{4.59a}$$

Since we have equivalently

$$2^{-1} \int_{t}^{2\sqrt{a}} ds(4a - s^2)^{1/2} = \lambda \int_{r}^{r_2} p(r')dr \quad , \tag{4.59b}$$

the following relation must hold

$$2^{-1} \int_{-2\sqrt{a}}^{2\sqrt{a}} ds(4a - s^2)^{1/2} = \lambda \int_{r_1}^{r_2} p(r)dr \quad , \tag{4.60a}$$

which determines the so far undetermined constant a (or c). When the integral on the left is evaluated, this relation becomes

$$(n + 1/2)\pi \equiv a\pi = \lambda \int_{r_1}^{r_2} p(r)dr \quad , \tag{4.60b}$$

where we have set a = n + 1/2 in anticipation of a later result. Its form now looks identical with the Bohr-Sommerfeld quantization rule. Note, however, that n is not necessarily an integer as yet, although it will turn out to be so in the end when the boundary condition is imposed on the solution of the homogeneous differential equation (4.55).

With f so defined by (4.56a) and on changing the variable from r to t, the differential equation for a parabolic cylinder function follows from (4.55):

$$[d^2/dt^2 + (a - t^2/4)]W(t) = 0 \quad . \tag{4.61}$$

Therefore, two independent solutions A_1 and B_1 in $[0,r_2]$ may be given in terms of parabolic cylinder functions $D_n(\pm t)$ [4.37]:

$$A_1 = (2\pi)^{-1/4}[\Gamma(n + 1)t']^{-1/2}D_n(-t) \tag{4.62a}$$

$$B_1 = (2\pi)^{-1/4}[\Gamma(n + 1)/t']^{1/2}\Gamma(-n)\pi^{-1}[-\cos n\pi D_n(-t) + D_n(t)] \quad , \tag{4.62b}$$

where the prime on t means a differentiation with respect to r.

As in the case of a single turning point, the solution u_ℓ to (4.53) may be written in terms of A_1 and B_1:

$$u_\ell(r) = \alpha_\ell(r)A_1(r) + \beta_\ell(r)B_1(r) \quad , \tag{4.63}$$

where α_ℓ and β_ℓ are in general variable functions that may be determined in the same manner as in Sect.4.2. The same procedure in this case will yield a set of two coupled first-order equations for α_ℓ and β_ℓ in exactly the same

form as in (4.31) with A and B replaced by A_1 and B_1 respectively. When
such an equation is solved, the wave function is obtained and consequently
the WKB solutions are corrected. Here we are not concerned with such a correction, being content with calculating the WKB phase shifts for the problem
in hand. In that case α_ℓ and β_ℓ may be regarded as constants. These constants,
however, must be such that the boundary conditions are properly satisfied.

The solution (4.63) in $[0,r_1]$ is joined by the solution in $[r_2,\infty]$ determined as follows. In the region $[r_2,\infty]$ (4.56b,57) give rise to the differential equation

$$[d^2/dt^2 + (t^2/4 - \alpha)]W(t) = 0 \quad , \tag{4.64}$$

where t now is related to r by the formula

$$2^{-1} \int_{2\sqrt{\alpha}}^{t} (s^2 - 4\alpha)^{1/2}ds = \lambda \int_{r_3}^{r} dr'p(r') \tag{4.65a}$$

or

$$2^{-1} \int_{t}^{2\sqrt{\alpha}} (4\alpha - s^2)^{1/2}ds = \lambda \int_{r}^{r_3} dr'|p(r')| \quad . \tag{4.65b}$$

On integration of the left part of (4.65b) from $-2\sqrt{\alpha}$ to $2\sqrt{\alpha}$ a relation for
α analogous to (4.60b) follows:

$$\pi\alpha = \lambda \int_{r_2}^{r_3} dr'|p(r')| \quad . \tag{4.66}$$

This relation shows that α is related to the Gamow factor [4.38]. The solutions to (4.64) are also parabolic cylinder functions $W(\alpha;\pm t)$ [2.12]. Therefore, the wave function u_ℓ in $[r_2,\infty]$ may be written as

$$u_\ell = \bar{\alpha}_\ell A_2 + \bar{\beta}_\ell B_2 \quad , \tag{4.67}$$

where

$$A_2 = (\pi t')^{-1/2}W(\alpha;t) \tag{4.68a}$$

$$B_2 = (\pi t')^{-1/2}W(\alpha;-t) \quad . \tag{4.68b}$$

Again, $\bar{\alpha}_\ell$ and $\bar{\beta}_\ell$ are, in general, variable functions of r that obey the same
equations as α_ℓ and β_ℓ (4.31), with A and B replaced by A_2 and B_2 respectively,
but we take them as constants as in the case of the wave function in $[0,r_1]$.
Then these constants $\bar{\alpha}_\ell$ and $\bar{\beta}_\ell$ can be determined in terms of the constants in
the region $[0,r_1]$ by matching logarithmically the two wave functions (4.63,
67) ar $r = r_2$. Thus we obtain

$$\bar{\alpha}_\ell = \pi([A_1,B_2]\alpha_\ell + [B_1,B_2]\beta_\ell) \tag{4.69a}$$

$$\bar{\beta}_\ell = -\pi([A_1,A_2]\alpha_\ell + [B_1,A_2]\beta_\ell) \quad , \tag{4.69b}$$

where $[X,Y]$ is the Wronskian of functions X and Y evaluated at $r = r_2$:

$$[X,Y] = [X(dY/dr) - Y(dX/dr)]_{r=r_2} \quad .$$

The asymptotic formulas for the parabolic cylinder functions $W(\alpha;\pm t)$ are

$$W(\alpha;t) \sim (2\kappa/t)^{1/2} \cos(t^2/4 - \alpha \log t + \pi/4 + \psi/2)$$

$$W(\alpha;-t) \sim (2/\kappa t)^{1/2} \sin(t^2/4 - \alpha \log t + \pi/4 + \psi/2) \tag{4.70}$$

where

$$\kappa = [1 + \exp(2\pi\alpha)]^{1/2} - \exp(\pi\alpha)$$

$$\psi = \arg \Gamma(1/2 + i\alpha) \quad . \tag{4.71}$$

Note that as $r \to \infty$

$$t^2/4 - \alpha \log t - 2^{-1} \alpha \log(e/\alpha) = \delta_\ell + kr - \ell\pi/2 \quad , \tag{4.72}$$

where

$$\delta_\ell = \int_{r_3}^\infty dr[\lambda p(r) - k] + \pi(\ell + 1/2)/2 - kr_3 \quad . \tag{4.73}$$

When (4.69-73) are used in the asymptotic form of u_ℓ, the following relationship arises:

$$u_\ell(r) \sim (\pi k)^{-1/2}[W^{(+)} \exp(ikr - i\ell\pi/2) + W^{(-)} \exp(-ikr + i\ell\pi/2)] \quad , \tag{4.74}$$

where

$$\gamma_\ell = 2^{-1} \alpha \log(e/\alpha) + \psi/2 + \pi/4 \tag{4.75}$$

$$W^{(\pm)} = \alpha^{-1/2}(\kappa^{1/2}\bar{\alpha}_\ell \mp i\kappa^{-1/2}\bar{\beta}_\ell)\exp[\pm i(\gamma_\ell + \delta_\ell)] \quad . \tag{4.76}$$

When (4.74) is compared with (4.3), the phase shift η_ℓ is obtained, i.e.

$$\eta_\ell = \delta_\ell + \gamma_\ell + \tan^{-1}(\kappa\bar{\alpha}_\ell/\bar{\beta}_\ell) \quad . \tag{4.77}$$

Since $\bar{\alpha}_\ell$ and $\bar{\beta}_\ell$ are given in terms of α_ℓ and β_ℓ, (4.69), but the latter are not determined as yet, they must be determined to proceed further. We do it by considering the behavior of u_ℓ at $r = 0$. We therefore return to (4.62,63) and consider the asymptotic behavior of the functions involved. Let us define the following abbreviating symbols:

$$(r,r_1) = [2\pi\lambda|p(r)|]^{-1/2} \exp\left[\lambda \int_r^{r_1} dr' |p(r')|\right]$$

$$(r_1,r) = [2\pi\lambda|p(r)|]^{-1/2} \exp\left[-\lambda \int_r^{r_1} dr'|p(r')|\right] .$$

The asymptotic forms for A_1 and B_1 in the classically forbidden region are found to be (Appendix 1)

$$A_1 \sim A^{(-)}(r_1,r) + A^{(+)}(r,r_1)$$

$$B_1 \sim B^{(-)}(r_1,r) + B^{(+)}(r,r_1) ,$$

where

$$A^{(-)} = -i(2\pi)^{1/4} \exp(ia\pi)(a/e)^{a/2}[\Gamma(a + 1/2)]^{-1/2} ,$$

$$A^{(+)} = (2\pi)^{3/4}(e/a)^{a/2}[\Gamma(a + 1/2)]^{1/2} \cos(a\pi)/\pi , \qquad (4.78)$$

$$B^{(-)} = (2\pi)^{1/4}(a/e)^{a/2}[\Gamma(a + 1/2)]^{1/2}\Gamma(- a + 1/2) \times$$

$$[1 + i \exp(ia\pi) \sin a\pi]/\pi ,$$

$$B^{(+)} = -(2\pi)^{3/4}(e/a)^{a/2}[\Gamma(a + 1/2)]^{1/2} \sin(a\pi)/\pi .$$

Therefore as $r \to 0$

$$u_\ell \sim (\alpha_\ell A^{(-)} + \beta_\ell B^{(-)})(r_1,r) + (\alpha_\ell A^{(+)} + \beta_\ell B^{(+)})(r,r_1) . \qquad (4.79)$$

Since u_ℓ must be finite at $r = 0$ and (r,r_1) is the irregular solution, we require

$$\alpha_\ell A^{(+)} + \beta_\ell B^{(+)} = 0 . \qquad (4.80)$$

By substituting from (4.78), we find

$$\alpha_\ell/\beta_\ell = \tan(a\pi) . \qquad (4.81)$$

Substituting this into (4.69,77), we finally obtain the phase shift in the form

$$\eta_\ell = \delta_\ell + \gamma_\ell - \tan^{-1}\left\{\kappa \frac{[A_1 \tan(a\pi) + B_1,B_2]}{[A_1 \tan(a\pi) + B_1,A_2]}\right\} \qquad (4.82)$$

for which we used the additivity of the Wronskians. Note that the Wronskians are evaluated at $r = r_2$. This is the desired phase-shift formula for the potentials admitting three turning points. Note that the symbol [,] in (4.82) means the Wronskian of the functions involved.

The appearance of the $\tan(a\pi)$ factor has physical significance, since every time a becomes a half integer,

$$a = n + 1/2 , \quad (n = 0,1,2,...) , \qquad (4.83)$$

the phase shift changes discontinuously and the ratio α_ℓ/β_ℓ becomes infinite,
that is, $\beta_\ell = 0$. This means that u_ℓ is simply proportional to $D_n(-t)$ which
is the wave function for an oscillator if n is an integer — a Hermite polynomial with argument t in the present case. Therefore, the discontinuous
changes in n_ℓ are due to the formation of a bound state with the vibrational
quantum number given by the Bohr-Sommerfeld quantization condition (4.60b).
Note also that the discontinuous change is modulated by the factor κ which
is given in terms of the Gamow factor $\exp(-2\pi\alpha)$.

A phase-shift formula [4.34,39] similar to (4.82) may be obtained by
using the tracing rules given previously. Since uniform wave functions are
used in the present derivation — and they are more accurate than the primitive
WKB solutions — (4.82) can be shown to contain additional correction factors,
and so it should be generally more accurate.

4.4 Semiclassical Approximation of Scattering Amplitude

The scattering amplitude (4.4) may be calculated numerically with the phase-
shift formulas provided by the foregoing theories. The computation is rather
straightforward. Nevertheless, we consider the *Ford-Wheeler* approximation
method for $f_k(\theta)$ [4.4], since it furnishes us with valuable insights into
the physics of collision.

In this approximation method (I) the phase shifts are calculated with the
WKB phase-shift formula (4.14); (II) the Legendre polynomial $P_\ell(\cos\theta)$ is approximated by its asymptotic formula

$$P_\ell(\cos\theta) \approx \sqrt{2} \ [(\ell + 1/2)\pi \ \sin\theta]^{-1/2} \ \sin[(\ell + 1/2)\theta + \pi/4] \qquad (4.84)$$

which is sufficiently accurate if the angle is such that

$$\ell \ \sin\theta \gtrsim 1 \quad ;$$

and (III) the summation over ℓ in (4.4) is replaced by the integral over ℓ.
Since the semiclassical phase shifts are used and the quantum number ℓ is
treated as a continuous variable, the approximation certainly can be termed
semiclassical. We shall drop the superscript WKB from n_ℓ^{WKB} for notational
brevity in the following discussion.

Let us first note that the classical deflection function $\chi(\ell)$ is related
to the phase shift as follows:

$$\chi(\ell) = 2(dn_\ell/d\ell) \quad . \qquad (4.85)$$

We then define the positive parameter ℓ_θ by

$$\chi(\ell_\theta) = \pm\theta \quad,$$

where the + sign is for the net repulsive deflection and the - sign is for the net attractive deflection.

When the approximations listed above are made in (4.4), the integral for the scattering amplitude follows

$$f_k(\theta) \simeq k^{-1}(2\pi \sin\theta)^{-1/2} \int_0^\infty d\ell(\ell + 1/2)^{1/2}[\exp(iq_\ell^{(-)}) - \exp(iq_\ell^{(+)})] \quad, \quad (4.86)$$

where

$$q_\ell^{(\pm)} = 2\eta_\ell \pm (\ell + 1/2)\theta \pm 1/4\pi \quad. \tag{4.87}$$

Since the phases $q_\ell^{(\pm)}$ change rapidly as ℓ increases, the major contribution to the integral derives from the neighborhood of the stationary-phase point, if there is any. The stationary-phase point is determined by

$$2(d\eta_\ell/d\ell) \pm \theta = 0 \quad, \tag{4.88}$$

that is, $\chi(\ell_\theta) = \theta$ or $\chi(\ell_\theta) = -\theta$. The uniqueness of the trajectory demands that only one of these alternatives is possible as a stationary-phase condition. We may thus choose the - sign in (4.88). Then by following the usual procedure in the stationary-phase or saddle-point method [2.4-6], we find the integral in the form

$$f_k(\theta) \simeq k^{-1}[(\ell + 1/2)/2 \sin\theta(d^2\eta_\ell/d\ell^2)]_{\ell=\ell_\theta} \exp(iq_{\ell\theta}^{(-)} + 1/4i\pi) \tag{4.89}$$

in which we have neglected the + term in (4.86) that contributes negligibly compared to the - term. In view of the condition $\ell \sin\theta \gtrsim 1$ on $P_\ell(\cos\theta)$ the scattering angles must be such that $\sin\theta \gtrsim 1/\ell_\theta$ in (4.89) for it to be a reasonable approximation.

If there is more than one stationary-phase point for the integral in (4.86), there should be contributions similar to (4.89) from all the stationary-phase points. In that case we find the scattering amplitude in the form

$$f_k(\theta) \simeq k^{-1} \sum_j [(\ell + 1/2)/2\sin\theta(d^2\eta_\ell/d\ell^2)]_{\ell=\ell_{\theta j}}^{1/2} \exp(iq_{\ell_{\theta j}}^{(-)} + 1/4i\pi) \tag{4.90}$$

where the sum is over all the stationary-phase points. In this case the cross section generally shows an undulatory pattern arising from the interference of various branches in (4.90). Such undulatory structure can be a useful source of information on the interaction potential.

If (4.89) is employed to calculate the cross section, we find

$$\sigma(\theta) \approx k^{-2}[(\ell + 1/2)(2\sin\theta)^{-1}(d^2 n_\ell/d\ell^2)]_{\ell=\ell_\theta} \quad ,$$

which in view of (4.85) may be written

$$\sigma(\theta) = b/|d\chi/db|\sin\theta \quad . \tag{4.91}$$

Here b is the impact parameter defined by

$$b = (\ell_\theta + 1/2)\hbar/mv \quad .$$

Notice that (4.91) is precisely the classical form for the differential cross section.

If there is an attractive and a repulsive branch in the interaction potential, then the classical deflection function passes through a maximum. In this case the phases near the maximum play an important role and the integral in (4.86) takes a different form. Again, we retain only the negative phase term which contributes more, compared to the positive phase term.

The phase $q^{(-)}$ may be approximated as

$$g_\ell^{(-)} = q_0 + (\ell - \ell_r)(\theta_r - \theta) + q_3(\ell - \ell_r)^3/3 \tag{4.92}$$

where

$$q_0 = 2n_{\ell r} - \pi/4 - (\ell_r + 1/2)\theta$$

$$q_3 = (d^3 n_\ell/d\ell^3)_{\ell=\ell_r}$$

$$\theta_r = 2(dn_\ell/d\ell)_{\ell=\ell_r}$$

and θ_r is such that

$$(d^2 n_\ell/d\ell^2)_{\ell_r} = 0 \quad .$$

When (4.92) is substituted into (4.86) and the integral representation of the Airy function (2.22a) (with the contour taken along the real axis), the scattering amplitude follows in the form

$$f_k(\theta) \simeq k^{-1}[2\pi(\ell_r + 1/2)/\sin\theta]^{1/2} q_3^{-1/3} \exp(iq_0)\text{Ai}[q_3^{-1/3}(\theta_r - \theta)] \quad .$$

Analysis of the rainbow scattering cross-sections is an important source of information on the potential parameters, since the former can be measured fairly accurately in molecular beam experiments [4.21].

5. Inelastic Scattering: Coupled-State Approach

The semiclassical method for elastic scattering described in the previous chapter can be generalized to obtain the wave functions for inelastic transitions in a semiclassical approximation. We approach the problem from the viewpoint of the time-independent theory. The question then centers around how one might obtain the WKB wave functions for a system of coupled Schrö-dinger equations and construct the S matrix therefrom. There are a number of possible ways to go about it; they may be classified into two basically distinctive approaches. In one method the coupled Schrödinger equations are converted into a set of coupled (nonlinear) Riccati equations [5.1-5] gen-eralizing the Riccati equation arising in the phase-amplitude method for one-dimensional Schrödinger equations. Alternatively, they are converted into coupled integral equations for generalized phases [5.6,7]. In the other method, the coupled second-order equations are converted into coupled first-order linear differential equations. When they are exactly solved, both methods should give the same results, but since that is not the case, there may be numerical differences in their results. Although there has never been a numerical comparison made between the two approaches, the latter approach seems to have an advantage in that it is easier to deal with linear equations than nonlinear ones. For this reason and also for lack of space we shall consider only the linear differential equation approach. The theory is deve-loped stepwise by employing two different representations of potentials, that is, adiabatic and diabatic, for constructing the lowest-order WKB solu-tions. We compare the accuracy of these representations toward the end of the chapter.

5.1 Primitive Adiabatic WKB Wave Functions

Since we wish to study semiclassical approximations for inelastic scattering in close-coupling theory, we must first investigate a way to obtain asymptotic solutions for the coupled radial Schrödinger equation. To consider this question, it is sufficient to put the coupled equations, e.g., (3.38), in the form

$$[d^2/dr^2 + \lambda^2 E_i(r)]u_i(r) = \lambda^2 \sum_{j \neq i}^{n} w_{ij} u_j(r) \quad . \tag{5.1}$$

Here we assume that every quantity is dimensionless and λ is a large parameter defined by $\lambda = (2\mu\varepsilon)^{1/2} d/\hbar$, and the other symbols are defined by

$$E_i(r) = [\dot{E}_i - \ell_i(\ell_i + 1)/\lambda^2 r^2 - V_{ii}(r)]/\varepsilon$$

$$w_{ij}(r) = V_{ij}(r)/\varepsilon \quad (i \neq j) \quad ,$$

where ε and d are the mean strength and size parameter of $V_{ii}(r)$, respectively. We have compressed the quantum number set to a single index.

According to the well-established practice in the theory of differential equations, the coupled equations (5.1) may be written as a set of 2n coupled linear first-order differential equations, if a new set of vectors is defined as follows:

$$\phi_i = u_i \tag{5.2a}$$

$$\phi_{n+i} = \lambda(du_i/dr) \quad , \quad (i = 1,2,\ldots,n) \quad . \tag{5.2b}$$

The matrix form of the 2n coupled equations is then [5.8-10]

$$d\phi/dr = \lambda A(r)\phi(r) \quad , \tag{5.3}$$

where

$$A(r) = \begin{pmatrix} 0 & I \\ D & 0 \end{pmatrix} \tag{5.4}$$

$$D = \begin{pmatrix} -E_1(r) & w_{12} & w_{13} & \cdots & w_{1n} \\ w_{21} & -E_2(r) & w_{23} & \cdots & w_{2n} \\ \cdot & \cdot & \cdot & & \cdot \\ \cdot & \cdot & \cdot & & \cdot \\ \cdot & \cdot & \cdot & & \cdot \\ w_{n1} & w_{n2} & w_{n3} & \cdots & -E_n(r) \end{pmatrix} \tag{5.5}$$

and

$$\phi = \{\phi_1(r) \quad , \quad \phi_2(r) \, , \, \cdots \, , \quad \phi_{2n}(r)\} \quad . \tag{5.6}$$

Note that the matrix D is simply the matrix associated with the secular equations corresponding to the coupled equations (5.1).

To facilitate solving (5.3), we introduce an orthogonal transformation U such that

$$\phi(r) = U(r)\psi(r) \tag{5.7}$$

$$A(r) = U^{-1}(r)A(r)U(r) \quad . \tag{5.8}$$

Then, substitution of (5.7) into (5.3) yields a new set of equations

$$\frac{d}{dr}\,\psi(r) = [\lambda A(r) - \Gamma(r)]\psi(r) \quad , \tag{5.9}$$

where after some tedious calculations the matrix **A** is found to be in the following diagonal form,

$$A(r) = \begin{pmatrix} \mathbf{p} & 0 \\ 0 & -\mathbf{p} \end{pmatrix} \equiv \begin{pmatrix} i\Lambda & 0 \\ 0 & -i\Lambda \end{pmatrix} \tag{5.10}$$

with **p** denoting a diagonal matrix whose diagonal elements are the eigenvalues of secular matrix **D**

$$\mathbf{p} = \begin{pmatrix} p_1 & & 0 \\ & p_2 & \\ & & \ddots \\ 0 & & p_n \end{pmatrix} \quad , \quad (p_i = i\Lambda_i) \quad , \tag{5.11}$$

and the matrix **Γ** is given by

$$\Gamma(r) = U^{-1}(r)\,\frac{d}{dr}\,U(r) \quad . \tag{5.12}$$

It is noteworthy that the matrices $A(r)$ and $\Gamma(r)$ are both independent of the large parameter. Therefore the term $\lambda A(r)$ dominates the right-hand side of (5.9).

It can be shown that the orthogonal matrix **U** consists of the modal matrix **τ** for the eigenvalue problem of **D** and its eigenvalues:

$$U(r) = \begin{pmatrix} \tau & \tau \\ \tau\mathbf{p} & -\tau\mathbf{p} \end{pmatrix} \quad . \tag{5.13}$$

The modal matrix is constructed with the eigenvectors of **D**. If the column eigenvector corresponding to eigenvalue $p_i = i\Lambda_i$ is denoted by $e_{.i}$ ($i = 1,2, \ldots, n$), then the modal matrix is

$$\tau = (e_{.1}, e_{.2}, \ldots, e_{.n}) \tag{5.14}$$

and consequently

$$\tau\mathbf{p} = (p_1 e_{.1}, p_2 e_{.2}, \ldots, p_n e_{.n}) \quad , \tag{5.15}$$

where p_i multiplies the whole column vector $e_{.i}$.

Before proceeding further, it is useful to note the following various identities:

$$\det|\mathbf{U}| = (-2)^n \prod_{i=1}^{n} p_i = (-2i)^n \prod_{i=1}^{n} \Lambda_i$$

$$\mathbf{U}^{-1} = \frac{1}{2} \begin{pmatrix} \Delta & \mathbf{p}^{-1}\Delta \\ \Delta & -\mathbf{p}^{-1}\Delta \end{pmatrix} \tag{5.16}$$

$$\Delta = \begin{pmatrix} \Delta_{11} & \Delta_{21} & \cdots & \Delta_{n1} \\ \Delta_{12} & \Delta_{22} & \cdots & \Delta_{n2} \\ \vdots & \vdots & \ddots & \vdots \\ \Delta_{1n} & \Delta_{2n} & \cdots & \Delta_{nn} \end{pmatrix}$$

with Δ_{ij} being the (ij) cofactor of $\det|\tau|$ and

$$\sum_{k=1}^{n} \Delta_{kj}(de_{kj}/dr) = \sum_{k=1}^{n} \Delta_{jk}(de_{jk}/dr) = 0 \quad . \tag{5.17}$$

These identities are proved in [5.9,10] by using standard procedures of linear algebra.

With the help of these identities it is possible to show that the matrix Γ can be written in the form

$$\Gamma = \begin{pmatrix} \Gamma_d & \Gamma_0 \\ \Gamma_0 & \Gamma_d \end{pmatrix} \tag{5.18a}$$

where Γ_d and Γ_0 are $n \times n$ matrices with their elements defined by

$$\Gamma_{dij} = \gamma_i \delta_{ij} + P_{ij}^{(+)} \tag{5.18b}$$

$$\Gamma_{0ij} = -\gamma_i \delta_{ij} + P_{ij}^{(-)} \tag{5.18c}$$

$$\gamma_i = (2p_i)^{-1}(dp_i/dr) = (2\Lambda_i)^{-1}(d\Lambda_i/dr) \tag{5.18d}$$

$$P_{ij}^{(\pm)} - \frac{1}{2}(1 \pm p_j/p_i) \sum_{k=1}^{n} \Delta_{ki}(de_{kj}/dr) \quad . \tag{5.18c}$$

Therefore the diagonal components of the matrix Γ are simply

$$\Gamma_{ii} = \Gamma_{i+n,i+n} = (2\Lambda_i)^{-1}(d\Lambda_i/dr) \quad , \quad (i = 1,2,\ldots,n) \quad ; \tag{5.19}$$

an observation that will turn out to be useful for developing approximate solutions.

Since the first term on the right hand side of (5.9) is large compared
with the second, an approximate solution to (5.9) may be obtained by simply
neglecting the second. However, it is more useful to retain the diagonal
part of the second term especially in view of (5.19). Thus we define a set
of functions $\{\psi_i^{(0)}\}$ by

$$\frac{d}{dr} \psi^{(0)} = (\lambda A - \Gamma_{diag})\psi^{(0)} \quad , \tag{5.20}$$

where Γ_{diag} is the diagonal part of the matrix Γ. Since the 2n equations
above are now uncoupled, they are trivial to solve and we obtain

$$\psi_i^{(0)} = \Lambda_i^{-1/2} \exp\left[i\lambda \int^r dx \, \Lambda_i(x)\right] \tag{5.21}$$

$$\psi_{i+n}^{(0)} = \Lambda_i^{-1/2} \exp\left[-i\lambda \int^r dx \, \Lambda_i(x)\right] \, , \quad (i = 1,2,\ldots,n)$$

apart from the integration constants. When multiplied with the elements of
the matrix U, these approximate solutions in fact are the WKB solutions for
(5.1) in the adiabatic representation. Such WKB solutions were obtained for
two coupled Schrödinger equations by *Stückelberg* [5.11] in 1932 by a different method. In his method a fourth-order differential equation is obtained
by eliminating one equation out of two and then the equation is solved by
applying the usual WKB approximation method. The present method is much
simpler and more practical, especially if the number of coupled equations
is larger than 2.

The general WKB solutions for (5.1) may now be written as

$$u_i = \sum_{j=1}^{2n} B_j U_{ij} \psi_j^{(0)} \quad , \tag{5.22}$$

where B_j are the integration constants which can be determined if boundary
conditions are suitably imposed. This form of the general WKB solutions,
however, does not yield inelastic transition probabilities unless it is
used, for example, in the distorted-wave approximation [5.12] or any other
variation of it. To obtain inelastic transition probabilities within the
context of the WKB approximation, it is therefore necessary to examine
corrections to (5.22).

Such corrections may be achieved if the solutions to (5.9) are written as

$$\psi_i(r) = F_i(r)\psi_i^{(0)}(r) \quad , \quad (i = 1,2,\ldots,2n) \quad , \tag{5.23}$$

and the equations for $F_i(r)$ are obtained by substituting it into (5.9) and
making use of (5.21): in matrix form we obtain

$$\frac{d}{dr} F(r) = -M(r)F(r) \quad , \tag{5.24}$$

where the matrix **M** consists of the elements defined by

$$M_{ij}(r) = \Gamma_{ij}(r)\omega_{ji}(r)$$

$$M_{ii}(r) = 0 \tag{5.25}$$

$$\omega_{ji} = (\Lambda_i/\Lambda_j)^{1/2} \exp\left[i\lambda \int^r dx(\Lambda_j - \Lambda_i)\right] \quad ,$$

$$(i,j = 1,2,\ldots,2n) \quad .$$

The various elements of Γ are given in (5.18). It is useful to note that $\Gamma_{ij}(r)$ are nonoscillatory, whereas $\omega_{ij}(r)$ are oscillatory in the classically allowed region in the adiabatic representation. Equation (5.24) could be a starting point for a theory of inelastic scattering, but it would then require a more involved formulation than the equations to be discussed below. Since the main aim of this section has been to gain some useful insight into the semiclassical wave functions needed in the subsequent sections, we need not pursue the theory here. In the following sections we derive more suitable differential equations that yield corrections to the WKB solutions comparable to (4.27) by constructing the zeroth-order solutions equivalent to (5.21).

5.2 The Uniform Semiclassical Method

In the previous method the lowest-order solutions are given in terms of the primitive WKB solutions in the adiabatic representation $\psi_i^{(0)}$, which require careful treatment owing to the connection problem inherently associated with them, since the functions $\psi_i^{(0)}$ become singular at the turning points r_t where $\Lambda_i(r_t) = 0$. It is possible to eliminate such a weakness if we resort to the method of uniform semiclassical wave functions used in Sect.4.2. This method [5.13-15] discussed below is an extension of the method used in Sect. 4.2 for making corrections to the WKB phase shifts. In developing the theory for the present chapter we assume that the adiabatic potential functions admit single-turning points in the range of energy of interest. The case of multiple-turning points can be discussed by extending the method used in Sect.4.3 and will be discussed in a later chapter where semiclassical theory of predissociations will be presented.

Multi-channel theories are generally rather involved mathematically in any approach and the present theory is no exception. Therefore, we shall first discuss a two-state problem to show the gist of the idea and then generalize the results to an n-state problem. The theory develops as a generalization of the elastic-scattering theory (Sect.4.2) and hence parallels it in many aspects.

Assume that a set of functions $\{\phi_i(r); i = 1,2\}$ exists and form Schwartzian derivatives with them, i.e.,

$$\{\phi_i,r\} = (\phi_i'''/\phi_i') - \frac{3}{2} (\phi_i''/\phi_i')^2 \quad ,$$

where the primes denote differentiations with respect to r. Assuming that there is only one turning point for each potential and by a straightforward generalization of (4.16), the set of radial Schrödinger equations (5.1) may be written for n = 2 in the form [5.13-15],

$$\left[d^2/dr^2 + \lambda^2 \phi_i \phi_i'^2 + \frac{1}{2} \{\phi_i,r\} \right] u_i(r) = \sum_{j=1}^{2} g_{ij}(r,\lambda) u_j(r) \quad , \quad (i = 1,2)$$

(5.26)

where

$$g_{ij}(r,\lambda) = \lambda^2 [\phi_i \phi_i'^2 - E_i(r)] + \frac{1}{2} \{\phi_i,r\} \quad . \tag{5.27}$$

Note that the term $\phi_i \phi_i'^2$ is chosen since we are dealing with a single-turning-point situation. See (4.53,56) for a two-turning point case.

Let us define the following set of functions $Y_i(r)$ by the differential equations

$$\left[d^2/dr^2 + \lambda^2 \phi_i \phi_i'^2 + \frac{1}{2} \{\phi_i,r\} \right] Y_i(r) = 0 \quad . \tag{5.28}$$

When the transformation

$$Y_i = (\phi_i')^{-1/2} W_i(\phi_i)$$

$$t_i = \lambda^{2/3} \phi_i(r)$$

is introduced, (5.28) takes the form,

$$(d^2/dt_i^2 + t_i) W_i(t_i) = 0 \quad , \tag{5.29}$$

which is the Airy differential equation. The above equations have already appeared without the subscript i in Sect.4.2; see (4.20-22). We have not as yet specified $\phi_i(r)$ as a function of r: we shall do that presently. For the moment let us assume that they are well-defined functions of r.

We denote two independent solutions of (5.28) by A_j and B_j defined as follows:

$$A_j(r) = (t_j')^{-1/2} Ai(-t_j) \quad ,$$

$$B_j(r) = (t_j')^{-1/2} Bi(-t_j) \quad ,$$

(5.30)

which are also the subscripted versions of the functions in (4.26).

We now look for the solutions of (5.1) in the following form:

$$u_i(r) = \sum_{j=1}^{2} [\alpha_{ij}(r) A_j(r) + \beta_{ij}(r) B_j(r)] \quad ,$$

(5.31)

where $\alpha_{ij}(r)$ and $\beta_{ij}(r)$ are functions of r to be determined such that $u_i(r)$ are solutions to (5.1) subject to appropriate boundary conditions. Equation (5.31) is a generalization of (4.27). Since we have two states to consider, u_i is written as a linear combination of four independent functions A_j and B_j, (j = 1,2). Since there are 8 unknown functions for a system of two second-order differential equations, they are not all independent. Therefore, we may impose suitably chosen conditions on them. Since it is preferable to have the function α_{ij} and β_{ij} change as slowly as possible as a function of r, we demand that they satisfy the equations

$$\sum_{j=1}^{2} (\alpha_{ij}' A_j + \beta_{ij}' B_j) = 0 \quad , \quad (i = 1,2) \quad .$$

(5.32)

When u_i is differentiated with respect to r and condition (5.32) used, then

$$u_i' = \sum_{j=1}^{2} (\alpha_{ij} A_j' + \beta_{ij} B_j') \quad ,$$

(5.33)

where the prime means the derivative with respect to r. By imposing the following four additional conditions

$$[\psi_j - \lambda^2 E_i(r)] \alpha_{ii} + \lambda^2 w_{ij} \alpha_{ij} = 0 \quad ,$$

(5.34a)

$$[\psi_j - \lambda^2 E_i(r)] \beta_{ii} + \lambda^2 w_{ij} \beta_{ij} = 0$$

(5.34b)

with

$$\psi_j = \lambda^2 \phi_j \phi_j'^2 + \frac{1}{2} \{\phi_j, r\} \quad ,$$

(5.35)

differentiating (5.33) again with respect to r, and making use of (5.1,28), we obtain

$$\sum_{k=1}^{2} (\alpha_{ik}' A_k' + \beta_{ik}' B_k') = [(\psi_j - \lambda^2 E_i(r)) \alpha_{ii} + \lambda^2 w_{ij} \alpha_{ij}] A_i$$

$$+ [(\psi_j - \lambda^2 E_i(r)) \beta_{ii} + \lambda^2 w_{ij} \beta_{ij}] B_i \quad ,$$

(5.36)

where the free indices i and j are either 1 or 2 and $i \neq j$. This equation together with (5.32) is equivalent to (5.1).

It is useful for brevity of notation to define

$$\gamma_i = [\lambda^2 E_i(r) - \psi_i]/\lambda^2 w_{ij}$$

$$\dot{\delta}_i = [\psi_i - \lambda^2 E_j(r)]\gamma_i + \lambda^2 w_{ij} \quad . \qquad (i \neq j \; ; \quad i,j = 1,2) \qquad (5.37)$$

We also choose α_{ij} and β_{ij} ($i \neq j$) as dependent coefficient functions. Then (5.34) implies that

$$\alpha_{ij} = \gamma_i \alpha_{ii} \quad , \qquad \beta_{ij} = \gamma_i \beta_{ii} \quad , \qquad (i \neq j) \quad , \qquad (5.34c)$$

if $w_{ij} \neq 0$. We now eliminate α_{ij} and β_{ij} from (5.32,36) and obtain a set of linear coupled first-order differential equations for α_{ii} and β_{ii} ($i = 1,2$) which can be compactly written in matrix form as follows:

$$Q \frac{d}{dr} \Omega = M\Omega(r) \quad , \qquad (5.38)$$

where the column vector consists of the independent coefficient functions α_{ii} and β_{ii}

$$\Omega(r) = \{\alpha_{11}, \beta_{11}, \alpha_{22}, \beta_{22}\} \quad , \qquad (5.39)$$

and the matrices Q and M are given in terms of the Wronskian matrix of A_j and B_j,

$$W_j = \begin{pmatrix} A_j & B_j \\ A'_j & B'_j \end{pmatrix} \qquad (5.40)$$

among other quantities. They have the forms below:

$$Q = \begin{pmatrix} W_1 & \gamma_2 W_2 \\ \gamma_1 W_1 & W_2 \end{pmatrix} \qquad (5.41)$$

$$M = \begin{pmatrix} 0 & -\gamma'_2 W_2 + \delta_2 \\ -\gamma'_1 W_1 + \delta_1 & 0 \end{pmatrix} \qquad (5.42)$$

where the matrix δ_j is a 2×2 matrix defined by

$$\delta_j = \begin{pmatrix} 0 & 0 \\ \delta_j A_j & \delta_j B_j \end{pmatrix} \quad . \qquad (5.43)$$

Since with the conventional choice [2.12] of the normalization constants for the Airy functions the Wronskian of A_j and B_j is calculated to be

$\det |W_j| = $ Wronskian $(A_j, B_j) = 1/\pi$,

we easily find the determinant of Q:

$\det |Q| = [\pi^{-1}(1 - \gamma_1 \gamma_2)]^2$.

The inverse of Q is then found after some tedious calculations:

$$Q^{-1} = (1 - \gamma_1 \gamma_2)^{-1} \begin{pmatrix} W_1^{-1} & -\gamma_2 W_1^{-1} \\ -\gamma_1 W_2^{-1} & W_2^{-1} \end{pmatrix} . \tag{5.44}$$

When (5.38) is multiplied by this inverse matrix from left, a set of linear differential equations follows:

$$\frac{d}{dr} \Omega(r) = Q^{-1} M \Omega(r) . \tag{5.45}$$

Here the matrix product $Q^{-1}M$ takes a simple and interesting form if we choose ϕ_i suitably, as will be shown shortly.

There are two different ways to choose ϕ_i: one is in terms of adiabatic momenta and the other in terms of diabatic momenta. The former is preferable, since it not only gives better numerical results, especially in the strong-coupling regime, but also yields simpler equations than the latter, when the same type of approximation is applied. Therefore we use the former approach here. The latter is discussed in a subsequent section.

We now choose ϕ_i such that

$$\phi_i {\phi_i'}^2 = \frac{1}{2} \left(E_1(r) + E_2(r) + (-)^{i-1} \{ [E_1(r) - E_2(r)]^2 + 4w_{12}^2 \}^{1/2} \right)$$
$$\equiv \Lambda_i^2(r) , \qquad\qquad (i = 1,2) . \tag{5.46a}$$

The Λ_i is an eigenvalue of the secular matrix (5.5) for $n = 2$. Therefore, the study in Sect.5.1 to obtain approximate solutions provides us with a valuable clue for developing the theory here. The choice of ϕ_i above fixes the transformation from r to ϕ_i. By taking the square root of (5.46a) and integrating it, we obtain

$$\frac{2}{3} \phi_i^{3/2} = \int_{r_i}^{r} \Lambda_i(x)dx , \qquad (r > r_i) ;$$

$$\frac{2}{3} (-\phi_i)^{3/2} = \int_{r}^{r_i} |\Lambda_i(x)|dx , \qquad (r < r_i) , \tag{5.46b}$$

where r_i is the classical turning point of $\Lambda_i(r)$, i.e.,

$$\Lambda_i(r_i) = 0 \quad .$$

The transformation (5.46b) is the two-state version of (4.19). Note that ϕ_i are the variables of the Airy functions appearing in the present theory and the phase integrals are now given in terms of the adiabatic momentum Λ_i.

With the variables ϕ_i so chosen and with the definition

$$u = [E_1(r) - E_2(r)]/2w_{12} \quad ,$$

it is easy to show that

$$\gamma_i = (-)^{i-1}[u - (u^2 + 1)^{1/2}] + O(\lambda^{-2})$$

$$\delta_i = (-)^i(1 + u^2)^{1/2}\{\phi_i, r\} + O(\lambda^{-2}) \quad .$$
(5.47)

For brevity of equations to be presented let us define the following symbols:

$$P_{ij} = \begin{pmatrix} (A_i, B_j) & (B_i, B_j) \\ -(A_i, A_j) & -(B_i, A_j) \end{pmatrix}$$
(5.48)

$$(A_i, A_j) = \pi\left(A_i \frac{d}{dr} A_j - A_j \frac{d}{dr} A_i\right) \quad , \quad \text{etc.}$$
(5.49)

When the semiclassical limit (5.47) is taken, the matrix product $Q^{-1}M$ can be shown to take the form

$$M_\infty = \lim_{\lambda \to \infty} Q^{-1}M$$

$$= \frac{1}{2}[(u^2 + 1)^{1/2} - u](u^2 + 1)^{-1}\left(\frac{du}{dr}\right) I + \frac{1}{2}(u^2 + 1)^{-1}\left(\frac{du}{dr}\right) w$$

$$- \frac{1}{2}[(u^2 + 1)^{1/2} + u]w^{(1)}$$
(5.50)

where the matrices w and $w^{(1)}$ are defined by

$$W(u) = \begin{pmatrix} 0 & -P_{21} \\ P_{12} & 0 \end{pmatrix} \quad ,$$
(5.51)

$$w^{(1)} = \begin{pmatrix} \gamma_2\{\phi_1, r\}A_{11} & \{\phi_2, r\}A_{21} \\ -\{\phi_1, r\}A_{12} & -\gamma_1\{\phi_2, r\}A_{22} \end{pmatrix}$$

$$\mathbf{A}_{kj} = \pi \begin{pmatrix} A_k B_j & B_k B_j \\ -A_k A_j & -B_k A_j \end{pmatrix} \qquad (j,k = 1,2) \quad .$$

Finally, with the transformation

$$\Omega(r) = \{1 + [u - (u^2 + 1)^{1/2}]^2\}^{-1/2} X(u) \quad , \tag{5.52}$$

we obtain the set of linear first-order differential equations,

$$\frac{d}{du} X(u) = \frac{1}{2} (u^2 + 1)^{-1} \boldsymbol{w}(u) X(u) + \frac{1}{2} [(u^2 + 1)^{1/2} + u] \boldsymbol{w}^{(1)}(u) X(u) \quad . \tag{5.53}$$

The matrix $\boldsymbol{w}^{(1)}$ consists of the Schwartzian derivatives which indicate how fast the potential functions change over a distance. Therefore, in the case of slowly changing potentials the Schwartzian derivatives may be ignored and thus the matrix $\boldsymbol{w}^{(1)}$ may be dropped from the equation. In this case the equation takes a much simpler structure

$$\frac{d}{du} X(u) = \frac{1}{2} (u^2 + 1)^{-1} \boldsymbol{w}(u) X(u) \quad , \qquad (0 \leq u < \infty) \quad . \tag{5.53a}$$

This equation will be studied in subsequent discussions. With a further transformation

$$v = \tan^{-1} u \quad \left(-\frac{\pi}{2} \leq v \leq \frac{\pi}{2} \right) \quad ,$$

the differential equation may be written

$$\frac{d}{dv} X(v) = \frac{1}{2} \boldsymbol{w}(v) X(v) \quad . \tag{5.53b}$$

The solutions of these equations furnish the four independent variable coefficients α_{ii} and β_{ii} ($i = 1,2$). Since $\boldsymbol{w}$ consists of uniform semiclassical wave functions, (5.53) is valid everywhere at $0 \leq r < \infty$. This is in contrast to (5.24), which has singularities at the turning points that are not singularities of (5.1). We call (5.53) the equation of motion. The results obtained above and especially the equation of motion owe their relative simplicity to the limiting procedure taken for γ_i, δ_i and hence for $\mathbf{M}_\infty$, as well as the neglect of $\boldsymbol{w}^{(1)}$. The approximation required for (5.53a) is, in effect, mathematically equivalent to neglecting the Schwartzian derivatives $\{\phi_i, r\}$, which is $O(\lambda^0)$, in comparison with terms of $O(\lambda^2)$ in ψ_i. This is consistent with the idea of the WKB approximation, and in this sense (5.53a,b) in particular, thus scattering theory based thereon, is semiclassical.

By virtue of the limiting behavior (5.47), (5.34c) now takes the form

$$\alpha_{12} = - [u - (u^2 + 1)^{1/2}] \alpha_{22}$$

$$\alpha_{21} = [u - (u^2 + 1)^{1/2}]\alpha_{11}$$

and similarly for β_{12} and β_{21}. This implies that

$$\lim_{u \to \infty} \alpha_{ij} = 0$$

$$(i \neq j)$$

$$\lim_{u \to \infty} \beta_{ij} = 0$$

since α_{ii} and β_{ii} are finite in the limit $u \to \infty$, that is, in the limit of large distance.

Since the functions $A_j(r)$ tend to zero as r approaches 0, while the functions $B_j(r)$ increase exponentially, the boundary condition on $u_i(r)$ at $r = 0$ is satisfied if the initial value of $X(r)$ at $r = 0$ is such that

$$X_0 = \{\alpha_1^0, 0, \alpha_2^0, 0\} \quad . \tag{5.54}$$

When the equation of motion is solved subject to this condition, the wave functions u_i are obtained from which the scattering matrix can be constructed (Sect.5.3).

The matrix $\boldsymbol{w}$ is involutional [5.16-18] as are the matrices $\boldsymbol{w}$ and $\mathbf{F}_m$ in Sect.4.2. By making use of the Wronskian (5.50), it in fact is easy to verify that

$$\boldsymbol{w}^2 = -\mathbf{I} \quad .$$

Its integral is also involutional; let

$$\mathbf{F}_1 = \frac{1}{2} \int^v \boldsymbol{w}(v) \, dv \quad .$$

Then

$$\mathbf{F}_1^2 = -f(v)\mathbf{I}$$

$$f(v) = f_{11}f_{22} - f_{12}f_{21} \quad ,$$

where f_{ij} are the elements of the integral of the matrix $P_{12}/2$. These properties will prove very useful later for developing approximate solutions of the equation of motion (5.53).

The two-state theory above can be generalized to an n-state problem [5.14]. Here we assume again that the potentials admit only one turning point each. Under this assumption we introduce a set of transformations $\phi_i(r)$, $i = 1,2$ ...,n, whose precise forms will be specified similarly to (5.46b) presently, and we then form Schwartzian derivatives with $\phi_i(r)$.

Equation (5.26) now is easily generalized and the Schrödinger equation (5.1) may be written

64

$$\left[\frac{d^2}{dr^2} + \lambda^2\phi_i\phi_i^{\,\prime 2} + \frac{1}{2}\{\phi_i,r\}\right]u_i(r) = \sum_{j=1}^{n} g_{ij}(r;\lambda)u_j(r) \quad , \tag{5.55}$$

$$(i = 1,2,\ldots,n) \quad ,$$

where the notations are the same as in (5.26). As a generalization of (5.31), the wave functions are written

$$u_i(r) = \sum_{j=1}^{n} [\alpha_{ij}(r)A_j(r) + \beta_{ij}(r)B_j(r)] \tag{5.56}$$

with A_j and B_j defined by two independent solutions of (5.28), see (5.30).

Equation (5.56) contains $2n^2$ undetermined functions α_{ij} and β_{ij} and hence is underdetermined as it stands. We therefore impose on them conditions similar to those for the two-state theory just considered by generalizing (5.32):

$$\sum_{j=1}^{n} (\alpha_{ij}'A_j + \beta_{ij}'B_j) = 0 \cdot \quad , \quad (i = 1,2,\ldots,n) \quad . \tag{5.57}$$

The remaining $2n(n-1)$ conditions at our disposal are taken as a generalization of (5.34):

$$[\psi_j - \lambda^2 E_i(r)]\alpha_{ij} + \lambda^2 \sum_{\ell\neq i}^{n} w_{i\ell}\alpha_{\ell j} = 0$$

$$(i,j = 1,2,\ldots,n; \quad i \neq j) \;.$$

$$[\psi_j - \lambda^2 E_i(r)]\beta_{ij} + \lambda^2 \sum_{\ell\neq i}^{n} w_{i\ell}\beta_{\ell j} = 0 \tag{5.58}$$

Now following exactly the same procedure as for the case of $n = 2$ and making use of conditions (5.57,58) and equation (5.1), we obtain the first-order differential equations

$$\sum_{j=1}^{n} (\alpha_{ij}'A_j' + \beta_{ij}'B_j') = \left\{[\psi_i - \lambda^2 E_i(r)]\alpha_{ii} + \sum_{\ell\neq i}^{n} \lambda^2 w_{i\ell}\alpha_{\ell i}\right\}A_i$$

$$+ \left\{[\psi_i - \lambda^2 E_i(r)]\beta_{ii} + \sum_{\ell\neq i}^{n} \lambda^2 w_{i\ell}\beta_{\ell i}\right\}B_i \quad . \tag{5.59}$$

When (5.58) is used in (5.57,59), it is possible to eliminate α_{ij} and β_{ij} ($i \neq j$) in favor of α_{ii} and β_{ii} and thereby obtain a set of $2n$ coupled first-order differential equations for them. For this purpose it is convenient to introduce the following eigenvalue problem associated with the Schrödinger equation (5.1). We define an $n \times n$ secular matrix

$$F = [E_i(r)\delta_{ij} - w_{ij}(r)] \quad , \tag{5.60}$$

where the terms in the square brackets denote the matrix elements, and consider an eigenvalue problem

$$(F - \Lambda^2 I)h = 0 \quad , \tag{5.61}$$

where h is an n-dimensional eigenvector and Λ^2 the eigenvalue. Clearly, the secular matrix is associated with the Schrödinger equation (5.1), as can be seen easily if we make the following replacement in (5.1):

$$(d^2/dr^2) \rightarrow \Lambda^2 \quad .$$

We assume that the eigenvalue problem is solved. The eigenvalues are determined from the secular determinant

$$\det|F - \Lambda^2 I| = 0$$

and will be assumed all distinct and ordered suitably: Λ_1^2, Λ_2^2, ..., Λ_n^2. They are functions of r and $E_i(r)$. Since the coupling potentials vanish as $r \rightarrow \infty$, the secular determinant becomes diagonal at large distance and the eigenvalues tend to the channel wave numbers as follows:

$$\lim_{r \to \infty} \lambda \Lambda_i(r) = \lambda E_i^{1/2} \equiv k_i \quad . \tag{5.62}$$

This limit will be useful later.

The jth eigenvector will be denoted by column vector $h_{\cdot j} = \{h_{ij}; i = 1,2, ...,n\}$ which will be assumed orthonormal. The modal matrix may then be expressed by

$$H = (h_{\cdot 1}, h_{\cdot 2}, ..., h_{\cdot n})$$

which is an $n \times n$ matrix with the determinant equal to 1.

In accordance with the assumption that all potentials admit only one turning point each in the range of energy of interest, we assume that r_i exists such that

$$\Lambda_i(r_i) = 0 \quad . \tag{5.63}$$

In that case we again choose transformation functions $\phi_i(r)$ such that

$$\phi_i \phi_i'^2 = \Lambda_i^2(r) \quad , \quad (i = 1,2,...,n) \quad .$$

Note that the transformation functions are given by exactly the same relation as (5.46b) in terms of the adiabatic momentum Λ_i. Therefore, we see that in the semiclassical limit $\lambda \rightarrow \infty$

$$\psi_i = \lambda^2 \Lambda_i^2(r) \quad ,$$

which amounts to neglecting the Schwartzian derivative in ψ_i. To express α_{ij} in terms of α_{ii} and similarly for β_{ij}, it is useful to recast (5.58) in a slightly different form

$$[\psi_j - \lambda^2 E_i(r)]\alpha_{ij} + \sum_{\ell \neq i,j}^{n} \lambda^2 w_{i\ell}\alpha_{\ell j} = -\lambda^2 w_{ij}\alpha_{jj} \quad ,$$

and similarly·for the equation for β_{ij}. The coefficients of the column vector $\alpha_{\cdot j}$ in the above equation can be put into a matrix L_j which for $j = 1$, for example, has the form

$$L_1 = \begin{pmatrix} \varphi_{21} & \theta_{23} & \theta_{24} & \cdots & \theta_{2n} \\ \theta_{32} & \varphi_{31} & \theta_{34} & \cdots & \theta_{3n} \\ \vdots & \vdots & \vdots & & \vdots \\ \theta_{n2} & \theta_{n3} & \theta_{n4} & \cdots & \varphi_{n1} \end{pmatrix} \tag{5.64}$$

where $\varphi_{ij} = \psi_j - \lambda^2 E_i(r)$ and $\theta_{ij} = \lambda^2 w_{ij}(r)$. With these definitions and arranging L_j on the main diagonal, we are now able to write the whole set of equations for $\alpha_{\cdot j}$ in a more compact matrix form:

$$L\alpha = -\theta \quad , \tag{5.65}$$

where

$$\alpha = \{\alpha_{\cdot 1}, \alpha_{\cdot 2}, \cdots, \alpha_{\cdot n}\} \tag{5.66}$$

$$\theta = \{\theta_{\cdot 1}\alpha_{11}, \theta_{\cdot 2}\alpha_{22}, \cdots, \theta_{\cdot n}\alpha_{nn}\} \quad , \tag{5.67}$$

with $\alpha_{\cdot j}$ and $\theta_{\cdot j}$ denoting $(n-1)$-dimensional column vectors. The matrix L is by construction a block diagonal matrix with L_j on the main diagonal. A similar equation holds for β_{ij}. By solving (5.65) for α and utilizing the eigenvectors, it is possible to show that

$$\alpha_{ij} = (h_{ij}/h_{jj})\alpha_{jj} \tag{5.68a}$$

and similarly

$$\beta_{ij} = (h_{ij}/h_{jj})\beta_{jj} \tag{5.68b}$$

where $i, j = 1,2,\ldots,n$, $(i \neq j)$.

By substituting (5.68) into the right-hand side of (5.59) and taking the limit $\lambda \to \infty$, we find that it is equal to zero:

$$\lim_{\lambda \to \infty} \left\{ [\psi_i - \lambda^2 E_i(r)]\alpha_{ii} + \sum_{\ell \neq i}^{n} \lambda^2 w_{i\ell}\alpha_{\ell i} \right\} = 0 \quad ,$$

67

for which we have made use of the secular equation and $\psi_i = \lambda^2 \Lambda_i^2$ in the limit. The left-hand side of this equation is the generalization of the equation for δ_i in the two-state theory and it is $O(\lambda^{-2})$ when the Schwartzian derivative terms are neglected. If the Schwartzian derivative is disregarded on the left-hand side of the above equation in the limit $\lambda \to \infty$, it forms simply the secular equations (5.60) for a particular eigenvalue Λ_i^2 and thus the equality.

Finally, by combining (5.57,59) we obtain a set of 2n coupled first-order differential equations for α_{ii} and β_{ii} in the limit $\Lambda \to \infty$:

$$\frac{d}{dr} \, \Omega(r) = M_\infty \Omega(r) \tag{5.69}$$

where

$$\Omega(r) = \{\alpha_{11}, \, \beta_{11}, \, \alpha_{22}, \, \beta_{22}, \, \ldots, \, \alpha_{nn}, \, \beta_{nn}\} \tag{5.70}$$

$$M_\infty = M_{\infty d} + M_{\infty I} \tag{5.71}$$

with $M_{\infty d}$ denoting a diagonal matrix

$$M_{\infty d} = [(h_{ii}'/h_{ii})\delta_{ij}] \quad, \tag{5.72}$$

and $M_{\infty I}$ a traceless matrix consisting of two-dimensional matrices P_{ij} defined by (5.48),

$$M_{\infty I} = \begin{pmatrix} 0 & F_{12}P_{21} & \cdots & F_{1n}P_{n1} \\ F_{21}P_{12} & 0 & \cdots & F_{2n}P_{n2} \\ \vdots & \vdots & & \vdots \\ F_{n1}P_{1n} & F_{n2}P_{2n} & \cdots & 0 \end{pmatrix} \quad, \tag{5.73}$$

$$F_{ij} = (h_{ii}/h_{jj}) \sum_{k=1}^{n} H_{ki} h_{kj}' \quad. \tag{5.74}$$

Here H_{ki} is the (ki) cofactor of the determinant of the modal matrix H. It must be emphasized that (5.69) holds in the limit $\lambda \to \infty$ where terms of $O(\lambda^0)$ are neglected in comparison with terms of $O(\lambda^2)$ and in that sense it is a semiclassical equation. We call it the equation of motion. It is equivalent to (5.1).

Let us define a diagonal matrix $\Omega^{(0)}$

$$\Omega^{(0)}(r) = (h_{ii}\delta_{ij}) \tag{5.75}$$

and a new column vector $X(r)$ such that

$$\boldsymbol{\Omega}(r) = \boldsymbol{\Omega}^{(0)}(r)X(r) \quad . \tag{5.76}$$

On substitution of (5.76) into the equation of motion (5.69), the differential equation for $X(r)$ follows,

$$\frac{d}{dr} X(r) = \boldsymbol{w}(r)X(r) \quad , \tag{5.77}$$

where the matrix $\boldsymbol{w}(r)$ is similar to $M_{\infty I}$:

$$\boldsymbol{w}(r) = \boldsymbol{\Omega}^{(0)-1}M_{\infty I}\boldsymbol{\Omega}^{(0)} \quad . \tag{5.78}$$

This is the generalization of the equation of motion (5.53) in the two-state theory. The solution of (5.77) provides the desired scattering amplitudes in the present semiclassical theory. Since the wave functions u_i must all be equal to zero at $r = 0$, the initial condition on X is chosen as

$$X_0 = \{\alpha_1^0, 0, \alpha_2^0, 0, \ldots, \alpha_n^0, 0\} \tag{5.79}$$

so that the regular semiclassical wave function $A_j(r)$ determine the wave functions u_i at $r = 0$. Since the coupling potentials w_{ij} all vanish and therefore the secular determinant becomes diagonal at large distance, the components of the eigenvectors have the following limiting behavior:

$$\lim_{r \to \infty} h_{ij} = \delta_{ij} \tag{5.80}$$

and thus

$$\lim_{r \to \infty} F_{ij} = 0 \quad .$$

This implies in turn that the matrix $\boldsymbol{w}$ tends to zero as r increases and its rate is determined solely by the coupling potentials.

Methods for solving the equation of motion will be discussed in subsequent sections, as we develop approximations in terms of models of collision processes.

5.3 The Semiclassical Scattering Matrix Elements

Here we construct the semiclassical scattering-matrix elements for an n-state collision process from the solutions of the equation of motion (5.77). For the purpose it is sufficient to assume that the equation is somehow solved subject to the initial condition (5.79). Therefore, we denote the fundamental matrix for the equation by $a(r)$ and write the solution in the form

$$X(r) = \mathbf{a}(r)X_0 \cdot \quad , \tag{5.81}$$

so the solution of (5.69) becomes

$$\Omega(r) = \Omega^{(0)}\mathbf{a}(r)X_0 \quad . \tag{5.82}$$

The asymptotic form of $\Omega(r)$ is of interest for constructing the scattering-matrix elements. By using the definition of $\Omega^{(0)}(r)$, the limiting behavior (5.80), and the initial condition (5.79), we find

$$\Omega_i(\infty) = \lim_{r \to \infty} \Omega_i(r) = \sum_{j=1}^{n} a_{i,2j-1}\alpha_j^0 \tag{5.83}$$

where

$$a_{ik} = \lim_{r \to \infty} a_{ik}(r) \quad .$$

Furthermore, for $i \neq j$

$$\lim_{r \to \infty} \alpha_{ij} = \lim_{r \to \infty} \beta_{ij} = 0 \quad .$$

Since asymptotically

$$A_j(r) \sim (\pi k_j)^{-1/2} \sin(x_j + \pi/4)$$

$$B_j(r) \sim (\pi k_j)^{-1/2} \cos(x_j + \pi/4) \tag{5.84}$$

where

$$x_j = \lambda \int_{r_j}^{r} dx' \Lambda_j(x') \quad , \tag{5.85}$$

we obtain the asymptotic form for $u_i(r)$,

$$u_i(r) \sim (\pi k_i)^{-1/2}\Bigg[\Bigg(\sum_{j=1}^{n} a_{2i-1,2j-1}\alpha_j^0\Bigg)\sin(x_i + \pi/4)$$

$$+ \Bigg(\sum_{j=1}^{n} a_{2i,2j-1}\alpha_j^0\Bigg)\cos(x_i + \pi/4)\Bigg] \quad , \qquad (i = 1,2,\ldots,n) \quad . \tag{5.86}$$

Let us now designate state 1 as the initial state in which the scattering system is prepared. The boundary conditions consistent with it may then be taken as

$$u_1(r) \sim i^{\ell_1} \sin(k_1 r - \ell_1\pi/2) + T_{11} \exp(ik_1 r) \quad ,$$

$$u_j(r) \sim T_{1j} \exp(ik_j r) \quad , \quad (j = 2,3,\ldots,n) \quad . \tag{5.87}$$

When the asymptotic forms of the wave functions (5.86) are compared with these boundary conditions, a set of $2n$ linear equations for α_i^0 and T_{1j} follows:

70

$$T_{1j} + \delta_{j1}/2i = \frac{1}{2} i \exp(in_j)(\pi k_j)^{-1/2} \sum_{k=1}^{n} (a_{2j-1,2k-1} + ia_{2j,2k-1})\alpha_k^0$$

$$\tag{5.87a}$$

$$i^{\ell_j} \exp(in_j)(\pi k_j)^{1/2}\delta_{j1} = \sum_{k=1}^{n} (a_{2j-1,2k-1} - ia_{2j,2k-1})\alpha_k^0 \quad . \tag{5.87b}$$

The n parameters α_k^0 are determined from (5.87b). Solving these equations is equivalent to normalizing the initial probability amplitude. When α_k^0 so obtained are substituted in (5.87a), the scattering amplitudes T_{1j} are uniquely determined. The scattering-matrix elements are then determined as follows:

$$T_{1j} = (2i)^{-1}(k_1/k_j)^{1/2}(S_{1j} - \delta_{1j}) \quad , \tag{5.88}$$

where the scattering matrix is defined by the formula

$$S_{1j} = i^{\ell_1 - \ell_j}\Delta^{-1} \exp(in_1 + in_j)$$

$$\times \sum_{k=1}^{n} (a_{2j-1,2k-1} + ia_{2j,2k-1})\Delta_{1k} \quad , \tag{5.89}$$

with Δ_{1k} denoting the determinant of the linear system (5.87b)

$$\Delta = \det|a_{2j-1,2k-1} - ia_{2j,2k-1}| \quad , \tag{5.90}$$

Δ_{1k}, the (1,k) cofactor of the determinant, and n_j, the WKB phase shift

$$n_j = (\ell_j + 1/2)\pi/2 - k_j r_j + \int_{r_j}^{\infty} [\lambda\Lambda_j(x) - k_j]dx \quad .$$

Therefore, the problem is now reduced to finding the matrix a from the solutions of (5.77) subject to the initial condition (5.79). Its exact solution is not generally possible, but good approximate solutions are achievable which correlate well with experiment. The methods of approximation may be physically or mathematically motivated. Here, we simply point out that the equation of motion (5.77) may be cast in the same form as (4.39) by expressing X(r) in the exponential form

$$X(r) = \exp(S)X_0 \quad , \tag{5.91}$$

and then the resulting equation for S

$$\frac{d}{dr} S = \{S(\exp S - 1)^{-1}, \omega\} \tag{5.92}$$

may be solved iteratively [see (4.39) for the meaning of the curly commutator { , }]. Equation (5.77) may also be written as a product integral [4.26-28]

$$X(r) = \prod_{r_0}^{r} \exp[\mathbf{W}(s)ds]X(r_0) \ .$$ (5.93)

This form of the equation of motion is more useful than (5.92), if one has a numerical solution in mind.

As shown in Chap. 4, the first-order iterative solution is the first-order Magnus approximation [4.29-32] in which case

$$\mathbf{S} \simeq \int_{r_0}^{r} dx' \ \mathbf{W}(x') \ .$$ (5.94)

This form of solution is also obtained from (5.93) if

$$\mathbf{W}(s)\mathbf{W}(s') = \mathbf{W}(s')\mathbf{W}(s) \quad \text{for} \quad s \neq s' \ .$$

If the structure of the matrix $\mathbf{W}$ is such that it is involutional, the iterative series generated from (5.92) may be summable as discussed in Chap. 4. The details of some approximate solutions of (5.92) will be discussed in Chap. 7 in connection with applications of the theory. It must be pointed out that (5.94) gives fairly reasonable numerical results comparable to experiment for multi-channel scattering [5.19,20].

If (5.94) is taken for $\mathbf{S}$, then $a_{i,2j-1}$ appearing in the S-matrix elements are calculated by the formula

$$a_{i,2j-1} = \exp\left[\int_{r_0}^{\infty} dx \ \mathbf{W}(x)\right]_{i,2j-1} \ .$$ (5.95)

If (5.93) is used, then we have

$$a_{i,2j-1} = \left\{\prod_{r_0}^{\infty} \exp[\mathbf{W}(s)ds]\right\}_{i,2j-1} \ .$$ (5.96)

The form is very appealing for numerical computations of the S-matrix elements, since it can be directly computed with the matrix $\mathbf{W}(r)$. The exponential matrices are computed by expansion and, in practice, only a few leading terms are necessary for sufficiently accurate values of a_{ik}.

5.4 The Diabatic-Representation Approach

The uniform semiclassical method described in the previous section necessarily requires the adiabatic representation if the simplest set of coupled equations is sought for calculating scattering amplitudes. Although diagonalization of the secular equations is required to obtain the adiabatic-represen-

tation momenta, the uniform semiclassical wave functions appearing naturally
in the theory sufficiently compensate for the labor expended on the diagonalization, since they are free of the turning-point singularities. Since approximate theories generally result in some sort of expansion in the coupling parameters, the adiabatic-representation approach is probably more de-
sirable for strong coupling situations. This stems from the fact that the
adiabatic semiclassical wave functions are some kind of a resummation of an
expansion of diabatic semiclassical wave functions, as will be seen later
at the proper stage in the development of the theory. Nevertheless, the adiabatic representation is not the only representation for studying inelastic
scattering. Some authors [5.21-23] have formulated theories similar to that
in Sect.5.3 by means of functions of diabatic momenta. Here we consider this
alternative method for the sake of completeness and also because it can
yield good results when properly used, as will be shown later.

To simplify the presentation, we shall formulate the theory with the
example of a two-state problem and then present the n-state result without
proof, since by then the latter becomes almost self-evident especially in
view of the theory already presented in Sect.5.3.

Adiabatic and diabatic representation approaches differ in the choice
of the transformation functions $\phi_i(r)$. In the present case ϕ_i is chosen
such that

$$\phi_i \phi_i'^2 = E_i(r) \quad , \quad (i = 1,2) \quad . \tag{5.97}$$

Therefore, it is determined in terms of the diabatic momentum $p_i(r) = E_i^{1/2}(r)$.
Integrating (5.97), we obtain

$$\frac{2}{3} \phi_i^{3/2} = \int_{r_i}^{r} dr' \, p_i(r') \qquad (r > r_i)$$

$$\frac{2}{3} (-\phi_i)^{3/2} = \int_{r}^{r_i} dr' |p_i(r')| \quad (r < r_i) \quad , \tag{5.98}$$

where r_i is the classical turning point now defined by

$$p_i(r_i) = 0 \quad .$$

Note that (5.97) means

$$\psi_j = \lambda^2 p_i^2(r) + \frac{1}{2} \{\phi_i, r\}$$

which must be compared to (5.35).

Except for this difference the theoretical development in the diabatic
representation is completely parallel to the procedure used for deriving the
equation of motion (5.53 or 77). The wave functions are expanded as in (5.31)

and the conditions imposed on the coefficient functions are also formally the same as (5.32,34). The argument of the Airy functions is, of course, given by the function $\phi_i(r)$ defined now in (5.98). Despite the same formal structure, (5.34) yields γ_i and δ_i (5.37) quite different in form from those in the adiabatic semiclassical theory. To be more specific, in the latter approach γ_i tends to a finite limit independent of the large parameter λ, while δ_i tends to zero if the Schwartzian derivatives are neglected in the limit of large λ. But in the diabatic representation approach their limiting behaviors are interchanged, since

$$\lim_{\lambda \to \infty} \gamma_i = \lim_{\lambda \to \infty} \theta_i / 2\lambda^2 w_{12} = 0$$

$$\lim_{\lambda \to \infty} \delta_i = \lim_{\lambda \to \infty} \left(\lambda^2 w_{12} + \{(-)^{i-1}[E_1(r) - E_2(r)] + \theta_i / 2\lambda^2\}(\theta_i / 2w_{12}) \right) \quad (5.99)$$

$$\sim \lambda^2 w_{12} \quad ,$$

where θ_i are the Schwartzian derivatives: $\theta_i = \{\phi_i, r\}$. In any case, the procedure for deriving the equation of motion for the column vector composed of α_{ii} and β_{ii} is the same as for (5.45) and we obtain it in the form

$$\frac{d}{dr} \Omega(r) = M\Omega(r) \tag{5.100a}$$

with the matrix M defined by

$$M \equiv Q^{-1}\mathbf{M} = (1 - \gamma_1\gamma_2)^{-1} \begin{pmatrix} \gamma_2\gamma_1' I & \gamma_2' P_{21} \\ \gamma_1' P_{12} & \gamma_1\gamma_2' I \end{pmatrix}$$

$$+ (1 - \gamma_1\gamma_2)^{-1} \begin{pmatrix} -\gamma_2 W_1^{-1}\delta_1 & W_1^{-1}\delta_2 \\ W_2^{-1}\delta_1 & -\gamma_1 W_2^{-1}\delta_2 \end{pmatrix} \tag{5.100b}$$

where the prime denotes differentiation with respect to r and the matrix δ_j, $j = 1,2$, has the same form as (5.43) but with δ_j given by (5.99) this time.

We now examine the implication of the limiting behaviors of γ_i and δ_i and their role interchange as compared to (5.47): in the semiclassical limit of large λ the matrix M takes the limiting form

$$\lim_{\lambda \to \infty} M(\lambda) = \begin{pmatrix} 0 & W_1^{-1}\delta_2 \\ W_2^{-1}\delta_1 & 0 \end{pmatrix} + O(\lambda^{-2}) \equiv M_\infty + O(\lambda^{-2}) \quad . \tag{5.101}$$

Here we denote the matrix on the right-hand side by M_∞. When explicitly calculated, its elements are found to consist of products of the reference wave

74

functions A_j and B_j which are weighted by the coupling potential w_{ij}:

$$M_\infty = \lambda^2 w_{12} \begin{pmatrix} 0 & R_{21} \\ R_{12} & 0 \end{pmatrix} \equiv \lambda^2 w_{12} R \qquad (5.102)$$

$$R_{ij} = \pi \begin{pmatrix} -A_i B_j & -B_i B_j \\ A_i A_j & B_i A_j \end{pmatrix} \qquad (i,j = 1,2) \qquad (5.103)$$

where A_j and B_j are defined by (5.30) with ϕ_i now given by (5.98).

In the semiclassical limit the equation of motion (5.100a) takes the form

$$\frac{d}{dr} \Omega(r) = \lambda^2 w_{12} R(r) \Omega(r) \quad . \qquad (5.104)$$

This is the diabatic-representation-theory counterpart of (5.53). The equation of motion in this form furnishes us with a very interesting interpretation and insight into the theory itself. In order to see these more clearly, let us recall that the wave functions are given in the form

$$u_i = \sum_j [\alpha_{ij}(r) A_j(r) + \beta_{ij}(r) B_j(r)]$$

and the equation of motion (5.104) determines the coefficient functions α_{ij} and β_{ij} which are simply the corrections to the distorted wave functions A_j and B_j for the waves moving in the field of the diabatic potentials. There-fore, the elements of the matrix in (5.104) are related to the distorted-wave Born-matrix elements, since if (5.104) is integrated it is turned into an integral equation

$$\Omega(r) = \Omega(r_0) + \int_{r_0}^{r} dr' \, \lambda^2 w_{12}(r') R(r') \Omega(r') \quad ,$$

whose first iterative solution then consists of simply the distorted-wave Born matrix. Consequently, the equation of motion (5.104), when solved beyond first order, corrects the distorted-wave approximation, and the approximations developed in this section correspond to resummations of the distorted-wave Born expansion for the S matrix or transition probability amplitudes. Note that a similar interpretation applies to (5.53), since it is equivalent to (5.104). However, it should be recognized that since the adiabatic momenta are, so to speak, renormalized momenta and the Airy functions used in the adiabatic semiclassical theory are also resummed versions of $Ai(-\phi_j)$ and $Bi(-\phi_j)$ in the present approach, (5.53) may be regarded as a kind of resummed version of the equation of motion expanded in series of coupling potentials.

The solution of (5.104) supplies the desired scattering amplitudes. Similarly to (5.53), it is not possible to solve this equation analytically, but it is amenable to a variety of approximate-solution methods of good accuracy. For example, the first-order Magnus approximation is simple to work with, and can give quite good results. We also note here that M_∞ is involutional and hence the mathematical machinery developed in Chap. 4 for resumming the Magnus expansion is equally well applicable [5.24,25] with a considerable improvement in numerical accuracy of the transition probabilities. An application of this method is discussed in Chap. 7.

The foregoing method may be generalized to an n-state problem in a manner quite obvious by now. The wave functions are again written in terms of the reference functions A_j and B_j which are given by (5.30) with their argument now defined by (5.98). The correction functions are then determined such that their rate of change is as slow as possible, as expressed by the conditions (5.57). Conditions similar to (5.58) may also be imposed in terms of the diabatic momenta p_j,

$$(\psi_j - \lambda^2 p_i^2)\alpha_{ij} + \sum_{k \neq 1}^{n} \lambda^2 w_{ik}\alpha_{kj} = 0$$

$$(i,j = 1,2,\ldots,n; \quad i \neq j) \qquad (5.105)$$

$$(\psi_j - \lambda^2 p_i^2)\beta_{ij} + \sum_{k \neq 1}^{n} \lambda^2 w_{ik}\beta_{kj} = 0$$

where, because of the definition of ϕ_i,

$$\psi_j = \lambda^2 \phi_j \phi_j'^2 + \frac{1}{2}\theta_j = \lambda^2 p_j^2(r) + \frac{1}{2}\theta_j \qquad (5.106)$$

and thus it explains the appearance of the diabatic momenta in this approach. This set of equations may be solved to obtain the relations between the diagonal and off-diagonal components of the α and β coefficient functions which may be put in the form

$$\alpha_{ij}(r) = \gamma_{ij}(r)\alpha_{jj}(r)$$

$$\beta_{ij}(r) = \gamma_{ij}(r)\beta_{jj}(r) \quad , \qquad (5.107)$$

where γ_{ij} are determined by solving the linear sets (5.107). We assume that this is accomplished.

Now substituting these off-diagonal components into (5.57,59) and after some tedious manipulations, we obtain the equation of motion for the diagonal components of the coefficient functions. We shall not go into detail, but present the final results. Let us define the following:

$$\Omega(x) = \{\alpha_{11}, \beta_{11}, \alpha_{22}, \beta_{22}, \ldots, \alpha_{nn}, \beta_{nn}\}$$

$$\delta_i = \begin{pmatrix} 0 & 0 \\ \delta_i A_i & \delta_i B_i \end{pmatrix}$$

$$\delta_i = \frac{1}{2} \theta_i + \sum_{j \neq i} \lambda^2 w_{ij}(r) \gamma_{ji}(r)$$

$$Q(r) = \begin{pmatrix} W_1 & \gamma_{12}W_2 & & \gamma_{1n}W_n \\ \gamma_{21}W_1 & W_2 & \cdots & \gamma_{2n}W_n \\ \vdots & \vdots & & \vdots \\ \gamma_{n1}W_1 & \gamma_{n2}W_2 & \cdots & W_n \end{pmatrix}$$

$$P(r) = \begin{pmatrix} \delta_1 & -\gamma'_{12}W_2 & \cdots & -\gamma'_{1n}W_n \\ -\gamma'_{21}W_1 & \delta_2 & \cdots & -\gamma'_{2n}W_n \\ \vdots & \vdots & & \vdots \\ -\gamma'_{n1}W_1 & -\gamma'_{n2}W_2 & \cdots & \delta_n \end{pmatrix}$$

$$(\gamma'_{ij} = d\gamma_{ij}/dr)$$

where W_i are the Wronskian matrix for A_i and B_i whose argument is defined by (5.98); see (5.40). Then in matrix form the equation of motion is

$$Q \frac{d}{dr} \Omega(r) = P(r)\Omega(r) \quad . \tag{5.108}$$

By inverting the matrix Q, we get

$$\frac{d}{dr} \Omega(r) = M(r)\Omega(r) \quad , \tag{5.109}$$

where

$$M(r) = Q^{-1}(r)P(r) \quad . \tag{5.110}$$

This is the n-state generalization of (5.104). The more explicit form of the matrix M is

$$M = (\det|g|)^{-1} \begin{pmatrix} G_{11}W_1^{-1} & G_{21}W_1^{-1} & \cdots & G_{n1}W_1^{-1} \\ G_{12}W_2^{-1} & G_{22}W_2^{-1} & \cdots & G_{n2}W_2^{-1} \\ \vdots & \vdots & & \vdots \\ G_{1n}W_n^{-1} & G_{2n}W_n^{-1} & \cdots & G_{nn}W_n^{-1} \end{pmatrix} P(r)$$

where G_{ij} is the (i,j) cofactor of the determinant $\det|g|$ of the matrix $g(r)$,

$$g(r) = \begin{pmatrix} 1 & \gamma_{12} & \cdots & \gamma_{1n} \\ \gamma_{21} & 1 & \cdots & \gamma_{2n} \\ \vdots & \vdots & & \vdots \\ \gamma_{n1} & \gamma_{n2} & \cdots & 1 \end{pmatrix} .$$

In the semiclassical theory presented here we take the limit $\lambda \to \infty$ for the quantities involved in the equation of motion (5.109). Again by virtue of the imposed conditions on α_{ij} and β_{ij}, which are slightly different from those in the adiabatic-representation approach, the asymptotic behavior of γ_{ij} is different from the latter. In the present case of the n-state problem they tend to finite nonzero limits of $O(\lambda^0)$, while δ_i are still $O(\lambda^2)$. This implies that the matrix Q is $O(\lambda^2)$, while in the matrix P the diagonal components δ_i are $O(\lambda^2)$ and the off-diagonal components are $O(\lambda^0)$. Therefore in the large λ limit we may neglect the off-diagonal elements in P and approximate it by

$$P_\infty = \begin{pmatrix} \delta_1 & & & 0 \\ & \delta_2 & & \\ & & \ddots & \\ 0 & & & \delta_n \end{pmatrix} . \tag{5.111}$$

Then the large λ limit of M is

$$M_\infty = \lim_{\lambda \to \infty} M = Q^{-1}P_\infty$$

$$= (\det|g|)^{-1} \begin{pmatrix} G_{11}W_1^{-1}\delta_1 & G_{21}W_1^{-1}\delta_2 & \cdots & G_{n1}W_1^{-1}\delta_n \\ G_{12}W_2^{-1}\delta_1 & G_{22}W_2^{-1}\delta_2 & \cdots & G_{n2}W_2^{-1}\delta_n \\ \vdots & \vdots & & \vdots \\ G_{1n}W_n^{-1}\delta_1 & G_{2n}W_n^{-1}\delta_2 & \cdots & G_{nn}W_n^{-1}\delta_n \end{pmatrix} . \tag{5.112}$$

The large λ limit of the column vector $\Omega(r)$ is then determined by

$$\frac{d}{dr} \Omega(r) = M_\infty(r)\Omega(r) \quad , \tag{5.113}$$

which is the generalization of (5.104). This equation can be solved by a variety of methods, but since it has not been studied as extensively as its equivalent in the adiabatic-representation approach, it is not possible to comment on its numerical accuracy.

The solution of (5.109) or (5.113) supplies the matrix elements a_{ij} that comprise the S matrix. For example, the same solution methods can be applied to (5.109,113) as to (5.53,69) of the adiabatic representation theory, such as the first-order Magnus approximation and the product integral. The latter is especially suited for numerical integration methods. In any case the S-matrix elements can be directly computed from the product integral

$$a_{i,2j-1} = \left\{ \prod_{r_0}^{\infty} \exp[M(s)ds] \right\}_{i,2j-1} \tag{5.114}$$

which may be replaced by

$$a_{i,2j-1} = \left\{ \exp\left[\int_{r_0}^{\infty} ds\ M(s) \right] \right\}_{i,2j-1} \tag{5.115}$$

in the first-order Magnus approximation. Note that (5.114) is exact. It must be also noted that in diabatic S-matrix elements the phase shifts are now given in terms of the diabatic momentum,

$$\eta_j = \frac{1}{2} \pi (\ell_j + \frac{1}{2}) - k_j r_j + \int_{r_j}^{\infty} dr[\lambda p_j(r) - k] \quad .$$

Substitution of these results into (5.89) yields S_{1j} necessary for calculating the transition probabilities and cross sections for inelastic scattering.

6. Inelastic Scattering: Time-Dependent Approach

The scattering-matrix elements calculated in the previous chapters by means of time-independent methods can also be calculated by a time-dependent theory, as indicated in the preliminary discussions on scattering theories in Chap.3. Our purpose here is to calculate (3.20) or its generalization (3.32) in semiclassical approximations. There are many ad hoc methods available in the literature, devised to suit particular problems, e.g., the impact-parameter method. Here we wish to develop systematic semiclassical theories in which it is possible to see clearly to what order the terms are neglected. Although quite useful, the impact-parameter method does not seem to meet the criterion of this monograph and therefore will not be considered here. The reader is referred to many references for the method and its variations, of which only a handful is quoted here [3.1,2,6.1-10].

The most logical theory on which to build time-dependent semiclassical approximations is the *Feynman* formulation of quantum mechanics [6.11,12] and it is not surprising to see that the Feynman path-integrals have long been investigated from the semiclassical theory or correspondence principle point of view [6.13-24]. This Feynman approach is discussed below. Although very attractive and with advantageous features from the conceptual viewpoint, the Feynman path-integrals are difficult to calculate in practice, since they involve functional integrations. The semiclassical method is in fact regarded by some researchers as a way to perform functional integrations.

An alternative procedure is to use the idea originally put forth by *van Vleck* [6.25] and later discussed by *Dirac* [6.26]. This idea can be made to work for us to obtain well-defined semiclassical approximations [6.27,28] for the S-matrix elements [6.29]. Since it does not involve a functional integration method, it is easier to comprehend and the equations involved for solution also arise naturally and are quite clear. It will be discussed here as well.

We are concerned with molecular scattering in which some degrees of freedom of molecules execute periodic motions. In such a situation it is generally useful and convenient to formulate the theory in terms of action-angle variables. Therefore it would seem most expedient to develop semiclassical approximation methods entirely in terms of action-angle variables as was done by *Marcus* [6.29] and *Miller* [6.19]. However, such a formulation poses a difficulty when rigorously scrutinized, since the quantum-mechanical transformation from the Cartesian coordinates and conjugate momenta to the action-angle variables is not generally unitary, but a one-sided unitary transformation [6.30-34]. This one-sidedness of the unitary transformation gives rise to an undesirable result that the quantization rules in action-angle variables do not rigorously hold except for one dimension. Since the terms that destroy the rigorous quantization rules for action-angle variables are of the order of $\hbar$, one may ignore them in the semiclassical approximation as was done by *Marcus* [6.29]. However, the analysis in the previous chapters shows that the semiclassical scattering matrix gets contributions from terms of $O(\hbar)$. Therefore, it would be preferable not to neglect them from the outset. This means that it is necessary to formulate the theory without using the action-angle representation from the outset. To put these points in the proper perspective, we first consider unitary transformations in quantum mechanics and their correspondence to canonical transformations. Then, the semiclassical approximation for the S matrix will be considered along the line of the van Vleck-Dirac idea before the Feynman path method is examined. Since the semiclassical S matrix thus obtained requires some further evaluation of integrals, we shall also discuss asymptotic evaluations of the integrals involved, following *Stine* and *Marcus* [6.35], and *Connor* [6.36-39].

6.1 Canonical and Unitary Transformations

It is generally believed that those quantum-mechanical transformations corresponding to the canonical transformations of classical mechanics are unitary. *Dirac* showed in his celebrated book [6.26] on quantum mechanics that it is indeed the case when the transformation is from a set of operators (q,p) to another (Q,P). However, he considered only transformations from a set of operators with a continuous spectrum to another with the same spectrum. In this case the prescription is clear and without a hitch, and the transformation operators are unitary. On the other hand, if the new operators have a different spectrum from the old ones, it is not so clear

if the transformation is indeed unitary. Recent research [6.30-34] shows
that the transformations in that case are not unitary, but one-sided unitary.
According to *Moshinsky* and his collaborators [6.32-34], the basic reason for
that is that such canonical transformations are nonbijective — i.e., not one-
to-one — and thus necessitate the so-called ambiguity group and "ambiguity
spins" which label the ambiguity group. Although the analyses by *Moshinsky*
and his co-workers are mathematically tight and precise, they are also
rather elaborate and lengthy. Since our aim here is to make the essential
point, we follow *Leaf*'s argument [6.30,31] which precedes that of *Moshinsky*
et al.

Let us consider a canonical transformation from a one-dimensional Cartesian
coordinate q and its conjugate momentum p to the angle-action variable set
(Q,P) where the quantum-mechanical P now has a discrete spectrum. (Imagine
the situation of a harmonic oscillator). We assume that the discrete spectrum
is complete and the corresponding basis set is denoted by $\{|\mu>\}$. Thus we have

$$\sum_{\mu} |\mu><\mu| = I \quad , \quad <\mu'|\mu> = \delta_{\mu'\mu} \quad .$$

The transformation operator U transforms the basis $\{|q'>\}$ of the operator
with a continuous spectrum to the basis of an operator with a discrete spectrum. It is determined by

$$U^{\dagger}|q'> = \epsilon^{1/2} \sum_{\mu} \delta(q' - \mu)|\mu> \tag{6.1}$$

such that

$$U^{\dagger}U = I \quad . \tag{6.2}$$

The parameter ϵ is determined so that (6.2) is satisfied. Equation (6.2) then
may be written as

$$U^{\dagger}U = \epsilon \int dq' \sum_{\mu} \sum_{\mu'} \delta(q' - \mu)\delta(q' - \mu')|\mu><\mu'| \quad . \tag{6.3}$$

Since [6.40,41]

$$[\delta(q' - \mu)]^2 = \epsilon'\delta(q' - \mu)$$

where ϵ' is an arbitrary constant, if we choose $\epsilon = \epsilon'^{-1}$, there follows
from (6.3)

$$U^{\dagger}U = \epsilon\epsilon' \int dq' \sum_{\mu} \delta(q' - \mu)|\mu><\mu|$$

$$= \sum_{\mu} |\mu><\mu| = I \quad .$$

For the second line we have used the fact that $\varepsilon\varepsilon' = 1$ by the choice of ε'.
On the other hand, we obtain

$$UU^\dagger = \varepsilon \int dq' \int dq'' \sum_\mu \delta(q' - \mu)\delta(q'' - \mu)|q''><q'|$$

$$= \varepsilon \int dq'|q'> \sum_\mu \delta(q' - \mu)<q'|$$

$$= \varepsilon \sum_\mu \delta(q - \mu) \neq I \quad . \tag{6.4}$$

Equations (6.2,4) together show that the transformation U is not unitary:
it is a one-sided unitary transformation, if it transforms operators of a
continuous spectrum to those of a discrete spectrum.

Now let ε = infimum $|\mu - \mu'|$ where μ and μ' are two adjacent eigenvalues.
Then we may write in the limit of $\varepsilon \to 0$

$$\lim_{\varepsilon \to 0} \varepsilon \sum_\mu \delta(q - \mu) = \lim_{\varepsilon \to 0} \sum_\mu \int_{\mu - \frac{1}{2}\varepsilon}^{\mu + \frac{1}{2}\varepsilon} \delta(q - x)dx = I \quad .$$

That is,

$$\lim_{\varepsilon \to 0} UU^\dagger = I \quad ,$$

which means that the transformation U becomes unitary as the spectrum $\{\mu\}$
becomes continuous. *Moshinsky* et al. [6.32-34] obtained basically the same
conclusion, but with a more elaborate analysis.

Now let us consider what the above conclusion means for the semiclassical theories we are concerned with here. Consider a set of operators (q,p)
which transform to a new set (Q,P) under the transformation U

$$Q = U^\dagger qU$$

$$P = U^\dagger pU \quad . \tag{6.5}$$

Let us assume that (q,p) obey the canonical commutation relations,

$$[q_m,q_n] = [p_m,p_n] = 0 \quad ,$$

$$[q_m,p_n] = i\hbar\delta_{mn}I \quad . \tag{6.6}$$

If the transformation is unitary, i.e.,

$$U^\dagger U = UU^\dagger = I \quad , \tag{6.7}$$

then (6.6) implies the following canonical commutation relations for the
new operators,

$$[Q_m,Q_n] = [P_m,P_n] = 0 \quad ,$$

$$[Q_m, P_n] = i\hbar \delta_{mn} I \quad , \tag{6.8}$$

which is easy to derive from (6.6) by using (6.5,7).

Let us now assume that $U^\dagger U = I$, but $UU^\dagger \neq I$, as is the case when the transformation involves two operator sets with different spectra, one continuous and the other discrete. Let us assume that

$$UU^\dagger q = qUU^\dagger \tag{6.9}$$

but

$$UU^\dagger p \neq pUU^\dagger \quad . \tag{6.10}$$

This is easy to see, since if $UU^\dagger \neq I$, then it must be dependent on either q or p, or both. Therefore, if (6.9) holds, then (6.10) is implied. The reverse also holds. We then see

$$U^\dagger [q_m, p_n] U = i\hbar \delta_{mn} I \quad .$$

The left-hand side is now put in the form

$$U^\dagger [q_m, p_n] U = U^\dagger UU^\dagger q_m p_n U - U^\dagger p_n q_m UU^\dagger U$$

$$= U^\dagger q_m UU^\dagger p_n U - U^\dagger p_n UU^\dagger q_m U$$

$$= [Q_m, P_n] \quad .$$

Therefore, the commutation relation remains invariant. Similarly we obtain

$$[Q_m, Q_n] = 0 \quad ,$$

but we find

$$U^\dagger [p_m, p_n] U = U^\dagger p_m p_n U - U^\dagger p_n p_m U \neq 0$$

unless m = n.

In summary, if $U^\dagger U = I$, but $UU^\dagger \neq I$ and if $UU^\dagger q = qUU^\dagger$, but $UU^\dagger p \neq pUU^\dagger$, then

$$[Q_m, P_n] = i\hbar \delta_{mn} I$$

$$[Q_m, Q_n] = 0 \quad , \tag{6.11}$$

but

$$[P_m, P_n] \neq 0 \quad ,$$

if $\{Q, P\}$ is a new set of operators with spectra basically different from those of the old set of operators from which the former are obtained by the transformation U.

If we assume

$$UU^{\dagger}q \neq qUU^{\dagger}$$

$$UU^{\dagger}p = pUU^{\dagger} \quad ,$$

then we find

$$[Q_m, P_n] = i\hbar\delta_{mn}I \quad ,$$

$$[P_m, P_n] = 0 \quad ,$$

but

$$[Q_m, Q_n] \neq 0 \quad . \tag{6.12}$$

The canonical commutation relations hold for the set $\{Q,P\}$, i.e.,

$$[P_m, P_n] = 0 \quad \text{and} \quad [Q_m, Q_n] = 0 \quad ,$$

only in the case of one dimension. Therefore, we see that the usual quantization rules hold for action-angle variables only in one dimension, but if
there is more than one degree of freedom, the quantization rules in the
Cartesian coordinates do not remain invariant when transformed into action-
angle variable language. This has an important implication for the time-dependent semiclassical theories developed here, since it is necessary to develop the theories in Cartesian coordinates first and then transform the results into action-angle variable language.

6.2 Time-Dependent Semiclassical Theory

Time-dependent scattering theory may be used as a starting point for semiclassical approximation theories, if various operators and the S-matrix
operator in particular can be calculated in power series of the Planck constant $\hbar$. This program can be implemented by following *van Vleck* [6.25] and
Dirac [6.26] who showed that the wave functions can be expressed in terms
of classical action integrals. As is well known now, this is the starting
point of Feynman's formulation of quantum mechanics.

Since it is more convenient for presentation and for clarifying the
connection between canonical and unitary transformations (Sect.6.1), we develop the theory in stages, starting with simple elastic scattering and
building it up to more complicated situations.

The S matrix for a two-body scattering of particles without internal
degrees of freedom was seen to be given by

$$S = \lim_{\substack{t \to \infty \\ t_0 \to -\infty}} \exp(i\hbar^{-1}H_0 t) \exp[-i\hbar^{-1}H(t - t_0)] \exp(-i\hbar^{-1}H_0 t_0) \quad , \tag{6.13}$$

where H_0 and H are the unperturbed and total Hamiltonian. We assume that the unimportant center-of-mass Hamiltonian is factored out. Let **p** denote the momentum. The $|p\rangle$ is a simultaneous eigenvector of H_0 and **p**,

$$p|p\rangle = p|p\rangle$$

$$H_0|p\rangle = E|p\rangle \quad . \tag{6.14}$$

The matrix element of S in this basis set is

$$\langle p|S|p_0\rangle = \lim_{\substack{t \to \infty \\ t_0 \to -\infty}} \exp[i\hbar^{-1}E(t - t_0)]\langle p|\exp[-i\hbar^{-1}H(t - t_0)]|p_0\rangle \quad . \tag{6.15}$$

The second factor on the right-hand side of (6.15) is the evolution operator in the momentum representation,

$$T(p|p_0) = \langle p|\exp[-i\hbar^{-1}H(t - t_0)]|p_0\rangle \quad , \tag{6.16}$$

which obeys the Schrödinger equation in the momentum representation:

$$i\hbar \frac{\partial}{\partial t} T(p|p_0) = H(p,q)T(p|p_0) \quad , \tag{6.17}$$

where $p = p(t)$ and $p_0 = p(t_0)$, and **q** is of course given by $i\hbar(\partial/\partial p)$. Note that if **p** and **q** were taken for action and angle variables, **q** would not necessarily be $i\hbar(\partial/\partial p)$, as is obvious from (6.11,12). Therefore, if **(p,q)** are not in Cartesian coordinates, then they must be such that their eigenvalue spectra are the same as their counterpart in the Cartesian coordinate representation.

We may write

$$T(p|p_0) = \exp[i\hbar^{-1}\Xi(p,p_0)] \quad . \tag{6.18}$$

Since T is in general a complex-valued function, $\Xi(p,p_0)$ must be also complex. It is then convenient to write it in terms of two real functions $W(p,p_0)$ and $F(p,p_0)$:

$$\Xi(p,p_0) = W(p,p_0) + i\hbar F(p,p_0) \quad . \tag{6.19}$$

Then by substituting (6.18) into (6.17), we find

$$i\hbar \frac{\partial}{\partial t} A - A \frac{\partial W}{\partial t} = H(p,q - \nabla_p W)A \tag{6.20}$$

where

$$A = \exp[-F(p,p_0)] \tag{6.21}$$

and

$$\nabla_p = \partial/\partial \mathbf{p} \quad .$$

Note that we have made use of the identity

$$\exp(-i\hbar^{-1}W)H(\mathbf{q},\mathbf{p})\exp(i\hbar^{-1}W) = H[\mathbf{p},\exp(-i\hbar^{-1}W)\mathbf{q}\exp(i\hbar^{-1}W)]$$

$$= H(\mathbf{p},\mathbf{q} - \nabla_p W) \quad .$$

To facilitate solution of (6.20) we first expand F, W, and H into power series in $\hbar$ and compare terms of like power and separate real and imaginary parts to obtain a hierarchy of equations. Let

$$F = \sum_{n=0}^{\infty} \hbar^n F^{(n)} \quad ,$$

$$W = \sum_{n=0}^{\infty} \hbar^n W^{(n)} \quad , \qquad\qquad (6.22)$$

$$H = \sum_{n=0}^{\infty} \hbar^n H^{(n)} \quad .$$

Then we first find that

$$H^{(n)} = \left(\frac{\partial^n H}{\partial \hbar^n}\right)_{\hbar=0} = \left(\frac{i}{2}\right)^n \sum_{k=0}^{n} \binom{n}{k} \nabla_p^k \left(\frac{\partial^n H}{\partial \mathbf{q}^n}\right)_{\mathbf{q}=-\nabla_p W^{(0)}} \nabla_p^{n-k} \quad . \qquad (6.23)$$

For example,

$$H^{(0)} = H(\mathbf{p}, -\nabla_p W^{(0)})$$

$$H^{(1)} = \frac{i}{2}\left[\nabla_p \left(\frac{\partial H}{\partial \mathbf{q}}\right)_{\mathbf{q}=-\nabla_p W^{(0)}} + \left(\frac{\partial H}{\partial \mathbf{q}}\right)_{\mathbf{q}=-\nabla_p W^{(0)}} \nabla_p\right] \quad .$$

Thus by substituting (6.22) into (6.20), we obtain the hierarchy of equations

$$\frac{\partial W^{(0)}}{\partial t} + H(\mathbf{p},-\nabla_p W^{(0)}) = 0 \quad , \qquad\qquad (6.24)$$

$$\frac{\partial F^{(0)}}{\partial t} - \left(\frac{\partial H}{\partial \mathbf{q}}\right)_{\mathbf{q}=-\nabla_p W^{(0)}} \frac{\partial F^{(0)}}{\partial p} = -\frac{1}{2}\frac{\partial}{\partial p}\left(\frac{\partial H}{\partial \mathbf{q}}\right)_{\mathbf{q}=-\nabla_p W^{(0)}} \, , \, \cdots \quad . \qquad (6.25)$$

The first equation is the Hamiltonian-Jacobi equation [6.42] for the generating function $W^{(0)}(\mathbf{p},\mathbf{p}_0)$ such that

$$q_i = -\frac{\partial W^{(0)}}{\partial p_i} \quad , \qquad q_{i0} = \frac{\partial W^{(0)}}{\partial p_{i0}} \quad . \qquad\qquad (6.26)$$

To interpret the second equation, first define

$$A^{(0)} = \exp(-F^{(0)})$$

and note that by Hamilton's canonical equations of motion

$$\left(\frac{\partial H}{\partial \mathbf{q}}\right)_{\mathbf{q}=-\nabla_{\mathbf{p}}W^{(0)}} = - \; \frac{d\mathbf{p}}{dt} \quad . \tag{6.27}$$

Then (6.25) may be cast in the form

$$\frac{\partial A^{(0)2}}{\partial t} + \frac{\partial}{\partial \mathbf{p}} \cdot \left(\frac{d\mathbf{p}}{dt} A^{(0)2}\right) = 0 \quad . \tag{6.25a}$$

This is the equation of continuity in momentum space for $A^{(0)2}$ which may be regarded as the probability fluid density in momentum space.

In view of (6.26) the complete integral of (6.24) may be given in terms of the Lagrangian $L(\mathbf{q},\dot{\mathbf{q}})$ [6.43]:

$$W^{(0)}(\mathbf{p},\mathbf{p}_0) = \int_{t_0}^{t} dt \; L(\mathbf{q},\dot{\mathbf{q}}) + \sum_i (p_{i0}q_{i0} - p_iq_i) \quad . \tag{6.28}$$

Here it is useful to note that

$$W_1^{(0)}(\mathbf{q},\mathbf{q}_0) \equiv \int_{t_0}^{t} dt \; L(\mathbf{q},\dot{\mathbf{q}})$$

is a generating function — a complete integral — for the Hamilton-Jacobi equation with $\mathbf{q}$ and $\mathbf{q}_0$ taken as independent variables, and (6.28) is obtained from $W_1^{(0)}(\mathbf{q},\mathbf{q}_0)$ by the Legendre transformation from $(\mathbf{q},\mathbf{q}_0)$ to $(\mathbf{p}, \mathbf{p}_0)$.

The integral of (6.25) is obtained by using (6.27) and writing the left-hand side as

$$\frac{d}{dt} F^{(0)} = \frac{\partial F^{(0)}}{\partial t} + \frac{d\mathbf{p}}{dt} \cdot \frac{\partial F^{(0)}}{\partial t} \quad ,$$

which means that (6.25) may be integrated to

$$F^{(0)}(\mathbf{p},\mathbf{p}_0) = F_0^{(0)} + \frac{1}{2} \sum_i \int_{t_0}^{t} dt \; \frac{\partial}{\partial p_i} \left(\frac{dp_i}{dt}\right) \quad . \tag{6.29}$$

Here $F_0^{(0)}$ is the initial value of $F^{(0)}$. Equation (6.29) may be recast into another more suggestive form. For this purpose we write

$$\frac{dp_i}{dt} = \frac{\partial (p_1,p_2,\ldots,p_N)}{\partial (p_1,p_2,\ldots,p_{i-1},t,p_{i+1},\ldots,p_N)} \quad , \tag{6.30}$$

88

where N denotes the number of degrees of freedom and the right-hand side denotes a Jacobian. Let us define the following Jacobians

$$\Delta = \frac{\partial(q_{01},\ldots,q_{0N})}{\partial(p_1,\ldots,p_N)} \quad,$$

$$\Delta_i = \frac{\partial(q_{01},\ldots,q_{0N})}{\partial(p_1,\ldots,p_{i-1},t,p_{i+1},\ldots,p_N)} \quad.$$

Then (6.30) may be written in the form

$$\frac{dp_i}{dt} = \Delta_i/\Delta \quad.$$

Since the Jacobians defined above may be regarded as the cofactor of $\partial E/\partial t$, $\partial E/\partial p_1,\ldots,\ \partial E/\partial p_N$ in the Jacobian

$$\frac{\partial(E,q_{10},q_{20},\ldots,q_{N0})}{\partial(t,p_1,p_2,\ldots,p_N)} \quad,$$

the cofactors satisfy, by Jacobi's lemma [6.44][1], the relation

$$\sum_i (\partial\Delta_i/\partial p_i) = 0 \quad, \quad (\Delta_0 \equiv \Delta \; ; \quad p_0 = t) \quad,$$

and it follows that

$$\sum_i \frac{\partial}{\partial p_i} \frac{dp_i}{dt} = \sum_i \frac{\partial}{\partial p_i} (\Delta_i/\Delta)$$

$$= \Delta^{-1} \sum_i \left[\frac{\partial\Delta_i}{\partial p_i} - \left(\frac{\Delta_i}{\Delta}\right)\frac{\partial\Delta}{\partial p_i}\right]$$

$$= -\Delta^{-1} \left(\frac{\partial\Delta}{\partial t} + \sum_{i\neq 0} \frac{dp_i}{dt}\frac{\partial\Delta}{\partial p_i}\right)$$

---

1 Jacobi's lemma: Let A_0, A_1, A_2,...,A_N denote cofactors of $\partial f/\partial x_0$, $\partial f/\partial x_1$, $\partial f/\partial x_2$,...,$\partial f/\partial x_N$ in the $(N + 1)$ dimensional determinant

$$\sum\left(\pm\frac{\partial f}{\partial x_0}\frac{\partial f_1}{\partial x_1}\frac{\partial f_2}{\partial x_2}\cdots\frac{\partial f_N}{\partial x_N}\right) \quad. \quad \text{Then } \sum_i(\partial A_i/\partial x_i) = 0.$$

This lemma is obvious for N = 1 and can be proved by induction for larger N's [Ref.6.44, p.230].

$$= - \Delta^{-1} \frac{d\Delta}{dt} \quad .$$

Substitution of this into (6.29) yields

$$F^{(0)} = F_0^{(0)} - \log\Delta^{1/2}$$

and hence

$$\exp(-F)_{\hbar=0} = \exp(-F^{(0)}) = A_0^{(0)} \Delta^{1/2} \quad , \tag{6.31}$$

where $A_0^{(0)} = \exp(-F_0^{(0)})$ is a constant.

Now by combining (6.28,31) with (6.19) and then (6.18), we obtain the S matrix to first order in $\hbar$:

$$\langle \mathbf{p} | S | \mathbf{p}_0 \rangle = \lim_{\substack{t \to \infty \\ t_0 \to -\infty}} A_0^{(0)} \Delta^{1/2} \exp[i\hbar^{-1} E(t - t_0)]$$

$$\times \exp\left\{ i\hbar^{-1} \left[\int_{t_0}^{t} dt \, L(\mathbf{q},\dot{\mathbf{q}}) + \sum_j (p_{j0}q_{j0} - p_j q_j) \right] \right\} \tag{6.32}$$

$$= \lim_{\substack{t \to \infty \\ t_0 \to -\infty}} A_0^{(0)} \Delta^{1/2} \exp\left\{ i\hbar^{-1} \left[\sum_j \int_{t_0}^{t} dt(p_j \dot{q}_j + p_{j0}q_{j0} - p_j q_j) \right] \right\} \quad .$$

We have used $H = \sum_i p_i \dot{q}_i - L(\mathbf{q},\dot{\mathbf{q}})$ for the second line. In the present case of two-body scattering it is possible to show[2]

$$\Delta^{1/2} = 1 \quad .$$

---

2 Since in (6.32) all the quantities involved are classical, it is possible to introduce a canonical transformation as required. Thus we may transform the variables to spherical coordinates and conjugate momenta. Then we see that in the case of centrosymmetric potentials

$$dp_\theta/dt = dp_\varphi/dt = 0$$

for the polar and azimuthal components of momentum, and thus we have

$$F^{(0)} = F^{(0)} + \frac{1}{2} \int_{t_0}^{t} dt \, \frac{\partial}{\partial p_r} (dp_r/dt)$$

$$= F_0^{(0)} + \frac{1}{2} \log[(dp_r/dt)/(dp_r/dt_0)] \quad .$$

Since as $t \to \infty$ and $t_0 \to -\infty$

$$dp_r/dt \quad , \quad dp_r/dt_0 \to 0 \quad ,$$

we see that $\lim\limits_{\substack{t \to \infty \\ t_0 \to -\infty}} F^{(0)} = F_0^{(0)}$ and therefore, $\Delta^{1/2} = 1$.

The unitarity of S then demands that

$$F_0^{(0)} = 1 \quad .$$

With these results the S matrix element may now be written in the familiar form

$$\langle p | S | p_0 \rangle = \langle p_r, \ell | S | p_r, \ell \rangle = \exp(2i\eta_\ell^{WKB}) \tag{6.33}$$

where

$$\eta_\ell^{WKB} = \frac{1}{2}\left(\ell + \frac{1}{2}\right)\pi - kr_0 + \int_{r_0}^{\infty} dr[k_r(r) - k]$$

$$k_r(r) = (2m/\hbar^2)^{1/2}[E - V(r) - (\ell + \frac{1}{2})^2\hbar^2/2mr^2]^{1/2} \quad , \tag{6.34}$$

with k denoting the wave number of the relative motion, ℓ the orbital angular momentum, and r_0 the classical turning point. Thus we have obtained the WKB phase shift. It must be noted that we have used the Langer modification in (6.34) in an ad hoc manner. In the present approximation scheme we should have ℓ instead of $(\ell + \frac{1}{2})$, but we have used the latter to remain uniform in the expressions presented.

The above method may be applied to obtain the semiclassical approximation to the multichannel S matrix (3.32b):

$$S_{ba} = \lim_{\substack{t \to \infty \\ t_0 \to -\infty}} \exp(i\hbar^{-1}H_b t)P_b \exp[-i\hbar^{-1}H(t - t_0)]P_a \exp(-i\hbar^{-1}H_a t_0) \quad . \tag{6.35}$$

Let us denote the relative momentum in channel a by $\mathbf{P}_a$ and the internal quantum numbers collectively by E_{ia}. Then the eigenvectors may be denoted by $|\mathbf{P}_a, E_{ia}\rangle$ which obeys

$$H_a |\mathbf{P}_a, E_{ia}\rangle = E|\mathbf{P}_a, E_{ia}\rangle \quad .$$

The S matrix elements in this representation are given by

$$\langle \mathbf{P}_b, E_{ib} | S_{ba} | \mathbf{P}_a, E_{ia} \rangle = \lim_{\substack{t \to \infty \\ t_0 \to \infty}} \exp[i\hbar^{-1}E(t - t_0)]T(\mathbf{P}_b, E_{ib} | \mathbf{P}_a, E_{ia}) \tag{6.36}$$

where

$$T(\mathbf{P}_b, E_{ib} | \mathbf{P}_a, E_{ia}) = \langle \mathbf{P}_b, E_{ib} | \exp[-i\hbar^{-1}H(t - t_0)] | \mathbf{P}_a, E_{ia}\rangle \quad . \tag{6.37}$$

This evolution operator satisfies the Schrödinger equation. *Marcus* [6.29] proposed to use the Schrödinger equation in the action-angle representation. In view of the discussion in Sect.5.1 we shall carry out the quantization in the Cartesian coordinate system or in a generalized coordinate system, in which the operators have the same spectra as those in the Cartesian coordinate system, e.g., the spherical coordinate system. Therefore, it is now necessary to write the evolution operator in the following form:

$$T(\mathbf{P}_b, E_{ib} | \mathbf{P}_a, E_{ia}) = \int d\mathbf{p}_a \int d\mathbf{p}_b \; <E_{ib} | \mathbf{p}_b>$$

$$\times <\mathbf{P}_b, \mathbf{p}_b | \exp[-i\hbar^{-1}H(t - t_0)] | \mathbf{P}_a, \mathbf{p}_a> <\mathbf{p}_a | E_{ia}> \quad . \tag{6.38}$$

The right-hand side may be expressed in terms of the evolution operator in the coordinate representation instead of the momentum representation. In either way the same result will be obtained. The evolution operator on the right-hand side of (6.38) satisfies the Schrödinger equation

$$i\hbar \frac{\partial}{\partial t} T(u_b | u_a) = H(u, i\hbar \nabla_u) T(u_b | u_a) \quad , \tag{6.39}$$

where the variables are abbreviated by collective symbols,

$$u_a = \{\mathbf{P}_a, \mathbf{p}_a\} \quad , \quad \text{etc.}$$

We shall sometimes denote the conjugate coordinates collectively by q_a.

Now we write $T(u_b | u_a)$ in terms of the functions W and F,

$$T(u_a | u_b) = \exp[i\hbar W(u_a, u_b) - F(u_a, u_b)] \quad , \tag{6.40}$$

and proceed in exactly the same manner as in the case for the two-body situation considered previously. When W, F, and H are expanded in power series in $\hbar$, the leading terms $W^{(0)}$ and $F^{(0)}$ respectively satisfy the Hamilton-Jacobi and the probability flux conservation equations (6.24,25,25a). Their integrals may be obtained similarly to (6.28,29) and we finally obtain

$$T(u_b | u_a) = A_0^{(0)} \Delta_{ba}^{1/2} \exp\{i\hbar^{-1}[\phi_{ba} - E(t - t_0)]\} \quad , \tag{6.41}$$

where

$$\Delta_{ba} = \partial(q_{a1}, \ldots, q_{aN}) / \partial(u_{b1}, \ldots, u_{bN}) \tag{6.42}$$

and

$$\phi_{ba} = \sum_i \int_{t_0}^{t} dt \; u_{bi} \dot{q}_{bi} + \sum_i [u_{ai}(t_0) q_{ai}(t_0) - u_{bi}(t) q_{bi}(t)] \quad . \tag{6.43}$$

The unitarity of the S matrix demands that

$$A_0^{(0)} = (2\pi i\hbar)^{-n/2} \quad ,$$

where n is the number of the internal degrees of freedom. Therefore, the semiclassical approximation to the S matrix element is given by

$$\langle P_b, E_{ib} | S_{ba} | P_a, E_{ia} \rangle = (2\pi i\hbar)^{-n/2} \int dp_a \int dp_b \langle E_{ib} | P_b \rangle_{WKB}$$

$$\times \Delta_{ba}^{1/2} \exp[i\hbar^{-1}\phi_{ba}] \langle P_a | E_{ia} \rangle_{WKB} \quad , \tag{6.44}$$

where $\langle P_a | E_{ia} \rangle_{WKB}$, etc., denote the WKB approximation of $\langle p_a | E_{ia} \rangle$, etc., the momentum representations of the internal wave functions.

This formula reduces to that in the classical S matrix theory [6.19] if

$$\langle P_a | E_{ia} \rangle = \delta[p_a - f(E_{ia})]$$

$$\langle P_b | E_{ib} \rangle = \delta[p_b - f(E_{ib})] \quad ,$$

where $f(E)$ is a well-behaved function of E. However, since this is not true generally, the classical S matrix

$$S_{ba}^{(c)} = (2\pi i\hbar)^{-n/2} \Delta_{ba}^{1/2} \exp(i\hbar\phi_{ba}) \tag{6.45}$$

is weighted by the semiclassical internal wave functions in the momentum representation.

Since the integrand in (6.44) oscillates, (6.44) may be evaluated by following the procedure developed by *Stine* and *Marcus* [6.35] and *Connor* [6.36-39], which we discuss below. The classical S matrix may be calculated numerically by solving the Hamilton-Jacobi or equivalent Hamilton's equations of motion subject to initial conditions (p_a, q_a) at $t_0 = -\infty$ which are consistent with the quantum number set $\{E_{ia}, P_a\}$. Those trajectories corresponding to the final state $\{E_{ib}, P_b\}$ are thus found. When $S_{ba}^{(c)}$ so obtained are averaged with the semiclassical weighting functions $\langle E_{ib} | P_a \rangle_{WKB}$ and $\langle P_a | E_{ia} \rangle_{WKB}$, the semiclassical S matrix (6.44) is obtained. This procedure avoids a double-ended initial conditions problem arising in (6.45), whose solution amounts to selecting phases (or coordinates) of a particular class demanded by the final state of the collision process of interest. However, it must be noted that classically the final phases and state are the outcome of a process for given initial conditions, but do not dictate that the system take a particular process meeting the double-ended initial conditions.

The momentum wave functions $\langle P_a | E_{ia} \rangle_{WKB}$ may be obtained by solving the Schrödinger equation for internal degrees of freedom in the momentum representation, or by Fourier-transforming the internal wave functions in coordinate representation into the momentum space.

An alternative form for the S matrix is given in the coordinate representation. It is easily obtained by taking the Fourier transform of (6.44) and changing

$$\Delta_{ba} \rightarrow \frac{\partial(q_1, \ldots, q_N)}{\partial(p_{01}, \ldots, p_{0N})} \equiv \overline{\Delta}_{ba}$$

and

$$\phi_{ba} \rightarrow \phi_{ba} + \sum_i [u_{bi}(t)q_{bi}(t) - u_{ai}(t_0)q_{ai}(t_0)] \equiv \Phi_{ba} \quad,$$

so that

$$\langle \mathbf{P}_b, E_{ib} | S_{ba} | \mathbf{P}_a, E_{ia} \rangle = (2\pi\hbar)^{-n/2} \int d\mathbf{q}_a \int d\mathbf{q}_b \langle E_{ib} | \mathbf{q}_b \rangle_{WKB}$$

$$\times \overline{\Delta}_{ba}^{-1/2} \exp[i\hbar^{-1}\Phi_{ba}] \langle \mathbf{q}_a | E_{ia} \rangle_{WKB} \quad, \qquad (6.46)$$

where the integrations are over all internal coordinates. The computational procedure for (6.46) is similar to that for (6.44). Since the semiclassical wave functions in (6.44,46) are singular at the classical turning points, it is advisable to employ uniformized semiclassical wave functions in actual computations.

6.3 The Feynman Path Integral Approach

Here we obtain (6.45) in an alternative way by using the Feynman path integral [6.11,12]. The propagator corresponding to (6.45) may be written as a path integral:

$$K(b,a) = \langle q_b | \exp[-i\hbar^{-1}H(t - t_0)] | q_a \rangle$$

$$= \int_a^b D[q(t)] \exp[i\hbar^{-1} \int_{t_0}^t dt \, L(q,\dot{q})] \quad. \qquad (6.47)$$

In the most recent theory [6.24] of Feynman path integrals, the path integrals such as (6.47) are written as integrals over the Wiener prodistribution and the semiclassical approximation to such an integral is obtained by expanding the exponent of the integrand around the classical path up to quadratic terms and then applying the *Cameron-Martin* formula [6.45]. This method is elegant and direct, if one is well versed in the integration theory in topological space. In spite of this advance, we shall rely on the naive approach commonly used, since it is more understandable to readers not versed in such mathematical techniques.

To evaluate (6.47) semiclassically, we follow the method first developed by *Morette* [6.13] and later by many others [6.15-20,27-29]. Let us denote the classical path by $\overline{q}_i$

$$q_i = \bar{q}_i + \lambda x_i \quad , \quad \lambda = (\hbar/m)^{1/2} \quad .$$

The classical path $\bar{q}_i$ is determined by the solution to the Hamilton-Jacobi equation

$$\frac{\partial S}{\partial t} + H\left(q_i , \frac{\partial S}{\partial q_i}\right) = 0 \quad , \tag{6.48}$$

where S is the canonical transformation function

$$S = S(q_i, \dot{q}_{i0}) \quad ,$$

such that

$$p_{i0} = -\partial S/\partial q_{i0}$$
$$p_i = \partial S/\partial q_i \quad . \tag{6.49}$$

Then the action is expanded in series of x_i to second order:

$$\begin{aligned} S &= \int_{t_0}^{t} dt\, L(\mathbf{q},\dot{\mathbf{q}}) \\ &= \bar{S} + (\lambda^2/2) \sum_{ij} \int_{t_0}^{t} dt[g_{ij}\dot{x}_i\dot{x}_j - V_{ij}x_ix_j] + O(\lambda^3) \quad , \end{aligned} \tag{6.50}$$

where

$$\bar{S} = \int_{t_0}^{t} dt\, L(\bar{\mathbf{q}},\dot{\bar{\mathbf{q}}}) \quad ,$$
$$g_{ij} = (\partial^2 L/\partial \dot{q}_i \partial \dot{q}_j)_{\mathbf{q}=\bar{\mathbf{q}}} \quad ,$$
$$V_{ij} = (\partial^2 L/\partial q_i \partial q_j)_{\mathbf{q}=\bar{\mathbf{q}}} \quad .$$

The first derivative does not appear since S is an extremum for the classical path. Substituting (6.50) into (6.47) yields the propagator in the form

$$K(b,a) = \exp[i\hbar^{-1}\bar{S}(\mathbf{q}_a,\mathbf{q}_b)] \int_{-\infty}^{\infty} D[x(t)]\exp\left[(i/2m) \sum_{ij} \int_{t_0}^{t} dt(g_{ij}\dot{x}_i\dot{x}_j - V_{ij}x_ix_j)\right] \quad . \tag{6.51}$$

The path integral in (6.51) is for a quadratic action and can be evaluated by following *Feynman* [6.12]. The integral does not depend on $\mathbf{q}_a$ and q_i, but on time t_0 and t. Therefore, by denoting the integral by $F(t,t_0)$, we may write the propagator in the form,

$$K(b,a) = F(t,t_0)\, \exp[i\hbar^{-1}\bar{S}(\mathbf{q}_a,\mathbf{q}_b)] \quad , \tag{6.52}$$

which is the classical approximation. To find $F(t,t_0)$ explicitly, we may use the unitarity of the propagator

$$\int d\mathbf{q} \; K(\mathbf{q}_b,\mathbf{q})K^*(\mathbf{q},\mathbf{q}_a) = \delta(\mathbf{q}_b - \mathbf{q}_a) \quad . \tag{6.53}$$

By substituting (6.52), applying the stationary-phase method for evaluating the integral thus obtained, and changing variables to $\{p_i\}$ and $\{q_i\}$, we finally find

$$F(t,t_0) = (2\pi i\hbar)^{-n/2} \left| \frac{\partial(q_{a1}, \; \cdots \; ,q_{aN})}{\partial(p_{b1}, \; \cdots \; ,p_{bN})} \right|^{1/2} \tag{6.54}$$

which when combined with (6.52) gives (6.45). Therefore, the present result can be grafted onto the semiclassical S matrix formula (6.44).

6.4 The Asymptotic Evaluation of the Semiclassical S Matrix

The S matrix elements obtained in the semiclassical approximation are given in terms of an integral which contains an oscillatory function in its integrand. Such integrals [6.35-39] may be evaluated by asymptotic methods especially in view of the fact that the exponent contains a large parameter that makes the integrand oscillate rapidly. Although there is still a lot to be learned about such multivariable integrals, there are some studies available [6.46-49] regarding the subject. I shall describe the essence of what is available, hoping to induce further developments.

The integral we wish to consider has the form

$$I(\alpha_1,\ldots,\alpha_m) = \int \ldots \int dx_1 \ldots dx_n g(x_1,\ldots,x_n,\alpha_1,\ldots,\alpha_m)$$

$$\times \exp[i\lambda f(x_1,\ldots,x_n,\alpha_1,\ldots,\alpha_m)] \quad , \tag{6.55}$$

where $(\alpha_1,\ldots,\alpha_m) \equiv \boldsymbol{\alpha}$ are parameters.

If there are sufficiently well-separated saddle points

$$\{x_i^{(j)} \; ; \; i = 1,2,\ldots,n\} \qquad (j = 1,2,\ldots)$$

defined by

$$\left(\frac{\partial f}{\partial x_i}\right)_{x_i=x_i^{(j)}} = 0 \tag{6.56}$$

and if

$$\Delta^{(k)} = \det \left| \left(\partial^2 f/\partial x_i \partial x_j\right)_{x_i=x_i^{(k)}, x_j=x_j^{(k)}} \right| \neq 0 \quad , \tag{6.57}$$

then (6.55) may be evaluated by means of the steepest-descent method. It yields the integral in the form

$$I \simeq \sum_k (2\pi i/\lambda)^{n/2} [\Delta^{(k)}]^{-1/2} \exp[i\lambda f(\boldsymbol{\alpha};\mathbf{x}^{(k)})] g(\boldsymbol{\alpha};\mathbf{x}^{(k)}) \quad . \tag{6.58}$$

This result is not uniform and will break down at the points where $\Delta^{(k)} = 0$. In some special cases such as one-dimensional integrals or integrals with nearly coinciding saddle points uniform asymptotic formulas can be obtained, but the theory is little developed at present for the general situation. *Connor* [6.36-39] has investigated some special cases by applying the method developed by *Ursell* et al. [6.46-48]. To get the basic idea of the method, let us discuss a two-dimensional integral with four nearly coinciding saddle points denoted by (x_i, y_i), $i = 1,2,3,4$. The integral of interest is

$$I = \int_c dx \int dy \, g(\boldsymbol{\alpha};x,y) \exp[i\lambda f(\boldsymbol{\alpha};x,y)] \quad , \tag{6.55a}$$

where λ is a large parameter and c denotes the contour.

In the case of four coalescing saddle points *Levinson*'s theorem [6.50] shows that the phase $f(\boldsymbol{\alpha};x,y)$ may be mapped into a cubic polynomial in (u,v):

$$f(\boldsymbol{\alpha};x,y) = u^3/3 + v^3/3 + \zeta(\boldsymbol{\alpha})u + \xi(\boldsymbol{\alpha})v + \eta(\boldsymbol{\alpha})uv + A(\boldsymbol{\alpha}) \quad . \tag{6.59}$$

The mapping is one-to-one if the saddle points (x_i, y_i), $i = 1,2,3,4$, correspond to their images (u_i, v_i), $i = 1,2,3,4$. That is,

$$(x_i, y_i) \to (u_i, v_i) \quad , \quad i = 1,2,3,4 \quad . \tag{6.60}$$

Equation (6.60) demands that the saddle points in (u,v) be determined by

$$\begin{aligned} u^2 + \eta v + \zeta &= 0 \\ v^2 + \eta u + \xi &= 0 \quad . \end{aligned} \tag{6.61}$$

Note that these are partial derivatives of $f(\boldsymbol{\alpha};x,y)$ with respect to u and v. The coefficients ζ, ξ, η, and A in (6.59) can be obtained by solving (6.61) to determine the saddle points (u_i, v_i) and substituting them into (6.59) to get four equations,

$$f^{(i)} = u_i^3/3 + v_i^3/3 + \zeta u_i + \xi v_i + \eta u_i v_i + A \quad , \tag{6.62}$$

where

$$f^{(i)} = f(\boldsymbol{\alpha};x_i, y_i) \quad , \quad i = 1,2,3,4 \quad .$$

With the mapping (6.59), integral I of (6.55a) may now be written as

$$I = \iint_{\Gamma} du\ dv\ g(\alpha;u,v)J(u,v)\ exp[i\lambda(u^3/3 + v^3/3 + \zeta u + \xi v + \eta uv + A)]\quad,$$

(6.63)

where $J(u,v)$ is the Jacobian and Γ denotes the contour in (u,v). Connor suggested that gJ be expanded as

$$gJ = p + qu + rv + suv\quad,$$

and eventually the exponential factor $exp(i\lambda\eta uv)$ be expanded into the Taylor series. We shall proceed somewhat differently.

First, we distort the contours in the u and v plane into the rays [u exp(i5π/6), u exp(iπ/6)] and [v exp(i5π/6), v exp(iπ/6)] whereby I is now given as

$$I = \int_{\infty\ exp(5i\pi/6)}^{\infty\ exp(i\pi/6)} du \int_{\infty\ exp(5i\pi/6)}^{\infty\ exp(i\pi/6)} dv\ g(\alpha;u,v)J(u,v)$$

$$\times\ exp[i\lambda(u^3/3 + v^3/3 + \zeta u + \xi v + \eta uv + A)]$$

(6.64a)

or taking the contours along the rays [u exp(iπ/3), u exp(-iπ/3)] and [v exp(iπ/3), v exp(-iπ/3)] by rotating the axes counter-clockwise by π/2

$$I = \int_{\infty\ exp(-i\pi/3)}^{\infty\ exp(i\pi/3)} du \int_{\infty\ exp(-i\pi/3)}^{\infty\ exp(i\pi/3} dv\ g(\alpha;iu,iv)J(iu,iv)$$

$$\times\ exp[\lambda(u^3/3 + v^3/3 - \zeta u - \xi v - i\eta uv + iA)]\quad.$$

(6.64b)

We now expand

$$g(\alpha;iu,iv)J(iu,iv)exp(-i\lambda\eta uv)$$

$$= \sum_{\ell m} [p_m(u^2 - \zeta)^m + q_m(u^2 - \zeta)^m u][r_\ell(v^2 - \xi)^\ell + s_\ell(v^2 - \xi)^\ell v]\quad, \quad (6.65)$$

where the coefficients p_m, etc., can be obtained by successively differentiating the left-hand side and evaluating the derivatives at $u = \zeta$ and $v = \xi$. Now substituting (6.65) into (6.64b) and performing integration by parts repeatedly after changes of variables to have the integrals defined on the real axis, we obtain for the asymptotic formula for I:

$$I = exp(i\lambda A)(2\pi i)^2 \prod_{k=1}^{2} \left[Ai(\lambda^{2/3}z_k) \sum_m \lambda^{-2m-1/3}A_m(z_k) \right.$$

$$+ Ai'(\lambda^{2/3}z_k) \sum_m \lambda^{-2m-5/3}B_m(z_k) + Ai'(\lambda^{2/3}z_k) \sum_m \lambda^{-2m-2/3}C_m(z_k)$$

$$\left. + Ai(\lambda^{2/3}z_k) \sum_m \lambda^{-2m-4/3}D_m(z_k) \right]\quad,$$

(6.66)

<u>Table 6.1.</u> Constants A_m, B_m, C_m, and D_m given in terms of the coefficients P_m, etc.

$A_0 = p_0$

$A_1 = 4p_3 + 12\zeta p_4 + 32p_5$

$B_0 = 2p_2$

$B_1 = 80p_5 + 120\zeta p_6$

$C_0 = -q_0$

$C_1 = -10q_3 - 12\zeta q_4$

$D_0 = -q_1 - 2\zeta q_2 - 4q_3$

$D_1 = 28q_4 - 140\zeta q_5 - (362 + 120\zeta^2)q_6 - 384\zeta^2 q_7$

---

where $z_1 = \zeta$, $z_2 = \xi$ and A_m, B_m, etc., are constants given in terms of p_m, q_m, r_ℓ and s_ℓ as well as ζ and ξ. The first few examples are given in Table 6.1. Equation (6.66) is an asymptotic expansion in the large parameter λ as shown by *Chester* et al. [6.46]. The following procedure is used to obtain (6.66). Let us first note that

$$Ai(z) = (2\pi i)^{-1} \int_{\infty \exp(-i\pi/3)}^{\infty \exp(i\pi/3)} du \, \exp(u^3/3 - zu)$$

and

$$Ai'(z) = -(2\pi i)^{-1} \int_{\infty \exp(-i\pi/3)}^{\infty \exp(i\pi/3)} du \, u \, \exp(u^3/3 - zu) \quad .$$

We then consider the integrals

$$I_p^{(m)} = \int_{\Gamma_u} du(u^2 - \zeta)^m \exp[\lambda(u^3/3 - \zeta u)] \quad . \tag{6.67}$$

$$I_q^{(m)} = \int_{\Gamma_u} du \, u(u^2 - \zeta)^m \exp[\lambda(u^3/3 - \zeta u)] \quad , \tag{6.68}$$

where Γ_u denotes the contour of integration.

If $m = 1$, we see that the integrals are, respectively, constant multiples of the Airy function and its derivative:

$$I_p^{(0)} = 2\pi i \lambda^{-1/3} Ai(\lambda^{2/3}\zeta)$$

$$I_q^{(0)} = -2\pi i \lambda^{-2/3} Ai'(\lambda^{2/3}\zeta) \quad .$$

If $m = 1$, we see that

$$I_p^{(1)} = 0 \quad ,$$

while

$$I_q^{(1)} = -2\pi i \lambda^{-4/3} Ai(\lambda^{2/3}\zeta) \quad .$$

This result follows from (6.68) on integration by parts and by use of the integral representation of the Airy function. Continuing in this manner, we may successively calculate the integrals (6.67,68) for all m.

The theory of asymptotic integration for integrals $I(\alpha_1,\ldots,\alpha_m)$ is not as yet fully developed for more general cases. It would, however, be very interesting and useful to study them.

7. Curve-Crossing Problems. I

Inelastic transition probabilities can be very much enhanced if there is a locally degenerate or almost degenerate state involved in the collision process. Such is the case with processes involving two or more potential curves crossing each other. Curve-crossing problems not only provide excellent models of inelastic transitions, but also facilitate well-defined and extremely fruitful analyses of inelastic-scattering processes in semiclassical approximations based on the theories discussed above. The aim of this chapter is manifold, since we wish to show the utility of the theories developed, by illustrating them with specific examples, and obtain a number of semianalytic formulas for the S matrix elements, which are all sufficiently accurate to be useful for practical analysis of experimental data on inelastic transitions in atomic and molecular systems.

In view of its historical interest and the insights it provides to inelastic transitions, we shall first consider the Landau-Zener-Stückelberg theory of two-state inelastic transitions [5.11,12,7.1]. Then we shall develop two-state theories based on the formalisms from the previous chapter. The two-state curve-crossing model is useful and important on its own merits, but it is also valuable since it is possible to build a multi-state curve-crossing theory with it, if the curve-crossing points are sufficiently well separated from each other. Such a multi-state theory is presented toward the end of the chapter.

7.1 The Landau-Zener-Stückelberg Theory

Applications of the WKB method ensued within a few years after its development in quantum mechanics. An important example is its application to collision problems and another is Gamow's theory of tunneling through a barrier by a particle. In 1932 *Landau* [5.12] used the WKB wave functions to calculate what is now known as a distorted-wave Born-matrix element. He recognized the importance of the curve-crossing point and its connection with stationary-phase points of the integral for the matrix element. Since the theory is based on the distorted-wave Born approximation, its applicability is limited to the range of small coupling constant. At about the same time *Zener* [7.1] and *Stückelberg* [5.11] examined the same problem with different approaches from Landau's. Zener took a time-dependent theory approach and obtained a more complete result by solving two coupled time-dependent (Schrödinger) equations in the variation-of-constant method. His result reduces to Landau's as the coupling constant decreases. On the other hand, Stückelberg took a time-independent theory approach and solved two coupled Schrödinger equations by the WKB method. The WKB solutions he obtained are in fact the same as those in (5.21). Therefore, his general solutions may be written as linear combinations of $\psi_i^{(0)}$, (i = 1,2,3,4):

$$ u_i = \sum_{j=1}^{4} U_{ij} \psi_j^{(0)} \qquad (i = 1,2) \quad , \qquad (7.1) $$

where the u_i's obey two coupled Schrödinger equations, e.g.,(5.1) with n = 2, and U_{ij} are the transformation matrix elements defined by (5.13) which become diagonal, i.e., $U_{ij} = \delta_{ij}$, asymptotically. This implies that (7.1) does not produce inelastic transition amplitudes. Note that this was the reason why it was necessary to proceed to a higher-order approximation than that for (5.21) and to develop theories described in Chap.5. Stückelberg recognized that the solutions (7.1) are not uniformly valid over the entire domain of the variable and that as asymptotic solutions they are subject to the Stokes phenomenon, when continued beyond the domain where crossing points, which turn out to be complex, are located. Thus, by applying the procedure developed by *Kramers* [2.11] and *Zwaan* [2.16], Stückelberg managed to produce inelastic transition amplitudes identical with Zener's. Thus, the celebrated Landau-Zener-Stückelberg theory came into being. Since the Stückelberg approach has a better heuristic aspect and, I believe, has more possibilities for application to more complicated problems, we shall consider it in detail in the following. For a critique of the Stückelberg theory, see [7.2].

For the two-state problem the eigenvectors $\mathbf{e}_{\cdot i}$ in (5.13-15) are given by

$$ e_{1i} = w_{12} \left\{ [E_1(r) - \Lambda_i^2]^2 + w_{12}^2 \right\}^{-1/2} $$

$$ e_{2i} = (-)^i [E_1(r) - \Lambda_i^2] \left\{ [E_1(r) - \Lambda_i^2]^2 + w_{12}^2 \right\}^{-1/2} \quad , \qquad (i = 1,2) \quad , \qquad (7.2) $$

where $E_i(r)$, i = 1,2, are defined in Sect.5.1 and

$$\Lambda_1 = 2^{-1/2}\left(E_1(r) + E_2(r) + \left\{\left[E_1(r) - E_2(r)\right]^2 + 4w_{12}^2\right\}^{1/2}\right)^{1/2}$$

$$\Lambda_2 = 2^{-1/2}\left(E_1(r) + E_2(r) - \left\{\left[E_1(r) - E_2(r)\right]^2 + 4w_{12}^2\right\}^{1/2}\right)^{1/2} \quad . \tag{7.3}$$

These are two roots of the secular determinant of the two-dimensional matrix **D** defined in (5.5). Then the WKB solutions $u_i(r)$ may be written as

$$u_i(r) = \sum_{j=1}^{2} \Lambda_j^{-1/2}\left\{B_j \exp\left[i\lambda \int_{r_j}^{r} dx' \, \Lambda_j(x')\right]\right.$$

$$\left. + B_{j+1} \exp\left[-i\lambda \int_{r_j}^{r} dx' \, \Lambda_j(x')\right]\right\}e_{ij}(r) \quad , \tag{7.4}$$

where B_j are constants and r_j are the classical turning points: $\Lambda_j(r_j) = 0$. It is convenient to define a new variable t by

$$t = [E_1(r) - E_2(r)]/2w_{12} \tag{7.5}$$

and the abbreviations similar to (2.31),

$$(r_j, r)_j = \Lambda_j^{-1/2} \exp\left[i\lambda \int_{r_j}^{r} dx' \, \Lambda_j(x')\right]$$

$$(r, r_j)_j = \Lambda_j^{-1/2} \exp\left[-i\lambda \int_{r_j}^{r} dx' \, \Lambda_j(x')\right] \quad . \tag{7.6}$$

The new variable just introduced makes it possible to remove the potentials from the formulas. The eigenvectors in particular take more illuminating forms in terms of t:

$$e_{11} = e_{22} = 2^{-1/2}(t^2 + 1)^{-1/4}[(t^2 + 1)^{1/2} + t]^{1/2} \quad ,$$

$$e_{12} = 2^{-1/2}(t^2 + 1)^{-1/4}[(t^2 + 1)^{1/2} - t]^{1/2} \quad ,$$

$$e_{21} = -e_{12} \quad .$$

The significance of these formulas will be discussed below. Note that it is also possible to write

$$e_{11} = \exp\left[-\frac{1}{2} \int^{t} dt' \left(\frac{t'}{t'^2 + 1} - \frac{1}{(t'^2 + 1)^{\frac{1}{2}}}\right)\right]$$

$$e_{12} = \exp\left[-\frac{1}{2} \int^{t} dt' \left(\frac{t'}{t'^2 + 1} + \frac{1}{(t'^2 + 1)^{\frac{1}{2}}}\right)\right] \quad . \tag{7.2b}$$

These expressions will turn out to be useful later on.

Since the WKB solutions (7.4) break down at the classical turning points r_j, a connection problem arises with them. In order to proceed further with this problem, we choose the phases of $\Lambda_j(r)$ such that

$$\Lambda_j(r) = -i|\Lambda_j(r)| \quad \text{for} \quad r < r_j \ .$$

Since the behavior of the wave functions $u_i(r)$ is basically determined by the exponential factors in the neighborhood of $r = 0$, the boundary conditions on $u_i(r)$ at $r = 0$ demand that B_2 and B_4 be equal to zero. Therefore the wave functions in the domain $0 < r < r_j$ $(j = 1,2)$ are in the form

$$u_i(r) = \sum_{j=1}^{2} B_j (r,r_j)_j \ e_{ij}(r) \ , \tag{7.7}$$

where

$$(r,r_j)_j = |\Lambda_j|^{-1/2} \exp\left[-\lambda \int_r^{r_j} dx' |\Lambda_j(x')| \right] \ . \tag{7.8}$$

The $(r_j,r)_j$ are defined with the (+) sign in the exponent. It is to be noted that $e_{ij}(r)$ are nonsingular functions of r in the neighborhood of r_j.

Now, by applying the rules of tracing presented in Chap.2, we obtain the solutions in the domain $r_j < r$, that is, in the classically allowed region:

$$u_i(r) = \sum_{j=1}^{2} \left[B_j^{(+)} \ e^{-i\pi/4} (r_j,r)_j + B_j^{(-)} \ e^{i\pi/4} (r,r_j)_j \right] e_{ij}(r) \ , \tag{7.9}$$

where $(r_j,r)_j$ and $(r,r_j)_j$ are now defined by (7.6) and

$$B_j^{(\pm)} = B_j/2\pi^{1/2} \ .$$

Put in other words, (7.9) is another form of the WKB connection formula in the classically allowed region which connects with the exponentially decreasing function in the classically forbidden region.

Let us assume that there is an avoided crossing point r_c sufficiently far from the turning points. Stückelberg realized that there is a pair of complex turning points associated with the avoided crossing point and the former divide the domain of the variable in essentially the same sense as a classical turning point does (Fig.7.1). That is, the WKB solutions are not uniformly valid across the point r_c. This means that it is necessary to investigate the connections of the solutions across the crossing point. Stückelberg found that the Zwaan-Kramers theory [2.11,16] is applicable. To implement this idea, it is necessary to recast (7.9) into a slightly different form

$$u_i(r) = \{r_c,r\}v_i + \{r,r_c\}v_i^* \ , \tag{7.10}$$

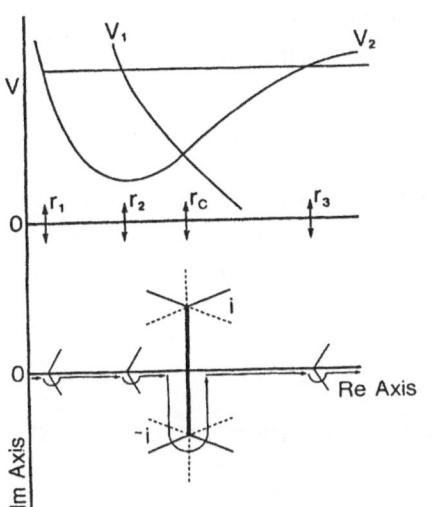

Tracing across the curve-crossing point. The curve-crossing point is r_c and r_1, r_2 and r_3 are the classical turning points. The lower panel shows the path of tracing

where

$$\{r_c, r\} = \exp\left[\frac{1}{2} i\lambda \int_{r_c}^{r} dx'(\Lambda_1 + \Lambda_2)\right] \tag{7.11a}$$

$$\{r, r_c\} = \exp\left[-\frac{1}{2} i\lambda \int_{r_c}^{r} dx'(\Lambda_1 + \Lambda_2)\right] \tag{7.11b}$$

$$v_i = \bar{B}_1^{(+)}(0,t)_{1i} + \bar{B}_2^{(+)}(t,0)_{2i} \tag{7.12a}$$

$$v_i^* = \bar{B}_1^{(-)}(t,0)_{1i} + \bar{B}_2^{(-)}(0,t)_{2i} \tag{7.12b}$$

$$(0,t)_{ij} = e_{ij}\Lambda_j^{-1/2} \exp\left[i \int_0^t dt'\, T(t')(t'^2 + 1)^{1/2}\right] \tag{7.13a}$$

$$(t,0)_{ij} = e_{ij}\Lambda_j^{-1/2} \exp\left[-i \int_0^t dt'\, T(t')(t'^2 + 1)^{1/2}\right] \tag{7.13b}$$

$$T(t) = \lambda w_{12}(\Lambda_1 + \Lambda_2)^{-1}(dr/dt) \tag{7.14}$$

$$\bar{B}_i^{(+)} = [r_i, r_c]B_i^{(+)} \equiv \exp\left[i\lambda \int_{r_i}^{r_c} \Lambda_i dx\right]B_i^{(')}$$

$$\bar{B}_i^{(-)} = [r_c, r_i]B_i^{(-)} \equiv \exp\left[-i\lambda \int_{r_i}^{r_c} \Lambda_i dx\right]B_i^{(-)} \quad .$$

Note that the exponents in (7.13a,b) may be expressed in another form:

$$\int_0^t dt'\, T(t')(t'^2 + 1)^{1/2} = \frac{1}{2} \lambda \int_{r_c}^{r} dx'[\Lambda_1(x') - \Lambda_2(x')] \quad . \tag{7.15}$$

It is assumed that $T(t)$ is a well-behaved function of t at $t = \pm i$. It is almost constant if the coupling potential w_{12} is a slowly changing function near the crossing point,

$$T(0) = T_0 \quad .$$

Then clearly, $T(t)(t^2 + 1)^{1/2}$ has zeros at $t = \pm i$ which act as if they were turning points. In fact, $(t,0)_{ij}$ and $(0,t)_{ij}$ may be regarded as the WKB solutions for a parabolic barrier problem, since in the vicinity of $t = \pm i$ it is possible to approximate e_{ij} by

$$e_{ij} \approx 2^{-1/2}(t^2 + 1)^{-1/4} \quad ,$$

and therefore the function $(0,t)_{ij}$, for example, is approximately given by

$$(0,t)_{ij} \approx (t^2 + 1)^{-1/4} \exp\left[i \int_0^t dt' \, T(t')(t'^2 + 1)^{1/2}\right]$$

$$\approx (t^2 + 1)^{-1/4} \exp\left[iT_0 \int_0^t dt'(t'^2 + 1)^{1/2}\right] \quad . \tag{7.16}$$

This is precisely a WKB solution to the Schrödinger equation for a parabolic barrier problem,

$$[d^2/dt^2 + T_0^2(1 + t^2)]\psi(t) = 0 \quad . \tag{7.17}$$

This equation is similar to the equation used by Zener [7.1] and it is not surprising that both theories give the same result for the transition probability. This implies that WKB solutions such as (7.16) must be traced around $t = \pm i$. Moreover, it implies that $(0,t)_{1i}$ and $(t,0)_{2i}$ would mix with each other, while $(t,0)_{1i}$ and $(0,t)_{2i}$ would also, when they are traced around the quasi-crossing points $t = \pm i$. This is the reason for writing u_i in the form of (7.10). Since v_i^* behaves in the complex conjugate sense to v_i, it is sufficient to treat only v_i to obtain the full result. The contour for tracing is defined in Fig.7.2. There are two sets of three Stokes lines issuing from $t = \pm i$ which are denoted by the solid lines, while the corresponding sets of three anti-Stokes lines are denoted by the broken lines. More precisely, the Stokes lines are defined by

$$Im\left\{\pm i \int_{\pm i}^t dt' \, T(t')(t'^2 + 1)^{1/2}\right\} = 0$$

and the anti-Stokes lines by

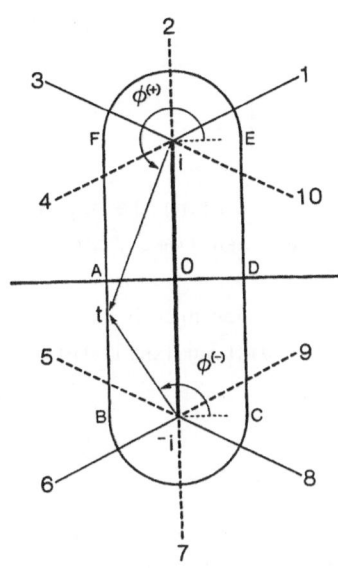

$$\text{Re}\left\{\pm i \int_{\pm i}^{t} dt'\ T(t')(t'^2 + 1)^{1/2}\right\} = 0 \quad .$$

The phases of the branches of $(t^2 + 1)^{1/2}$ are defined as follows:

$$\pi/2 < \phi^{(-)} < 5\pi/2$$

$$3\pi/2 < \phi^{(+)} < 7\pi/2 \quad ,$$

where

$$(t \pm i)^{1/2} = \rho_{\pm}\ \exp\left(\tfrac{1}{2}\ i\phi^{(\pm)}\right) \quad .$$

The Stokes constant associated with Stokes line ℓ is denoted by U_{ℓ}.

By dropping the subscript i from v_i for brevity of notation, we may write it in a form more suitable for initiating the tracing from Point A in Fig. 7.2:

$$v = C_1(-i,t)_1 + C_3(t,-i)_2 \tag{7.18}$$

where

$$C_1 = [0,-i]_1^{(-)} B_1^{(+)}$$

$$C_3 = [-i,0]_2^{(-)} B_2^{(+)} \quad , \tag{7.19}$$

with

$$[0,-i]_1^{(-)} = e_{1j}(-i)\ \exp\left[i \int_0^{-i} dt\ T(t)(t^2 + 1)^{1/2}\right]$$

107

$$[-i,0]_2^{(-)} = e_{2j}(-i) \exp\left[-i \int_0^{-i} dt \, T(t)(t^2 + 1)^{1/2}\right] \tag{7.20}$$

$$e_{kj}(-i) = (-)^{k-1} \exp\left[-\frac{1}{2} \int_0^{-i} dt \left(\frac{t}{t^2 + 1} - \frac{(-)^{j+k}}{(t^2 + 1)^{1/2}}\right)\right] . \tag{7.21}$$

The superscript (-) means that the integrals are performed along the path AB to the left of the branch cut in Fig.7.2. The sector between lines i and j is denoted by S(i,j) in Fig.7.2.

The branch $(-i,t)_1$ is dominant in S(A,5), while the other branch $(t,-i)_2$ is subdominant, the subscripts d and s on the branches distinguishing this difference:

$$v = C_1(-i,t)_{1d} + C_3(t,-i)_{2s} . \tag{7.22}$$

We are now ready for tracing.

When anti-Stokes line 5 is crossed into S(5,6), the terms in (7.22) exchange their dominancy and subdominancy according to one of the rules in Chap.2, i.e.,

$$v = C_1(-i,t)_{1s} + C_3(t_1,-i)_{2d} \quad \text{in} \quad S(5,6) . \tag{7.23}$$

When this is traced across line 6, which is a Stokes line, the dominancy and subdominancy are preserved, but according to the rules, the coefficients change as follows:

$$v = C_1^{(6)}(-i,t)_{1s} + C_3^{(6)}(t,-i)_{2d} \quad \text{in} \quad S(6,7) , \tag{7.24}$$

where

$$\begin{pmatrix} C_1^{(6)} \\ C_3^{(6)} \end{pmatrix} = \begin{pmatrix} 1 & U_6 \\ 0 & 1 \end{pmatrix} \begin{pmatrix} C_1 \\ C_3 \end{pmatrix} . \tag{7.25}$$

By continuing this process, we obtain the following forms for the solutions in various sectors:

$$v = C_1^{(6)}(-i,t)_{1d} + C_3^{(6)}(t,-i)_{2s} \quad \text{in} \quad S(7,8) \tag{7.26}$$

$$v = C_1^{(8)}(-i,t)_{1d} + C_3^{(8)}(t,-i)_{2s} \quad \text{in} \quad S(8,9) \tag{7.27}$$

$$v = C_1^{(8)}(-i,t)_{1s} + C_3^{(8)}(t,-i)_{2d} \quad \text{in} \quad S(9,D) , \tag{7.28}$$

where

$$\begin{pmatrix} c_1^{(8)} \\ c_3^{(8)} \end{pmatrix} = \begin{pmatrix} 1 & 0 \\ U_8 & 1 \end{pmatrix} \begin{pmatrix} c_1^{(6)} \\ c_3^{(6)} \end{pmatrix} . \tag{7.29}$$

When line 7 is crossed, (7.24) takes the form of (7.26) according to the rules since the line is an anti-Stokes line. Equation (7.27) arises from (7.26), when Stokes line 8 is crossed from $S(7,8)$ toward $S(8,9)$, according to the rules. Since line 9 is an anti-Stokes line, (7.28) arises from (7.27). When the solution (7.28) is extended beyond Point D and the reference point is switched from $-i$ to i, it takes the form

$$v = c_1^{(8)}[-i,i]_1^{(+)}(i,t)_{1d} + c_3^{(8)}[i,-i]_2^{(+)}(t,i)_{2s} \quad \text{in} \quad S(D,10) . \tag{7.30}$$

Here $[-i,i]_1^{(+)}$ and $[i,-i]_2^{(+)}$ are called dominancy-changing factors and are defined similarly to (7.20) with the integration being performed along the line CD. It must be noted that in (7.28,30) there is an exchange in dominancy between the branches. This can be understood as follows: the solution (7.28) can be extended across the positive real axis into the remote upper right-hand corner of Fig.7.1 and then as this solution crosses line 10 into $S(D,10)$, the dominancy exchange occurs according to the rule. Then, when the reference point is changed, (7.30) is obtained.

Now, the tracing in the upper half-plane of Fig.7.2 can be done exactly in the same manner as before, by starting from (7.30). To avoid repetition, we give the result only. The solution traced into $S(A,5)$ takes the form

$$v = c_1^{(A)}(-i,t)_{1d} + c_3^{(A)}(t,-i)_{2s} , \tag{7.31}$$

where

$$\begin{pmatrix} c_1^{(A)} \\ c_3^{(A)} \end{pmatrix} = \begin{pmatrix} [i,-i]_1^{(-)} & 0 \\ 0 & [-i,i]_2^{(-)} \end{pmatrix} \begin{pmatrix} 1 & 0 \\ U_3 & 1 \end{pmatrix} \begin{pmatrix} 1 & U_1 \\ 0 & 1 \end{pmatrix}$$

$$\times \begin{pmatrix} [-i,i]_1^{(+)} & 0 \\ 0 & [i,-i]_2^{(+)} \end{pmatrix} \begin{pmatrix} c_1^{(8)} \\ c_3^{(8)} \end{pmatrix} . \tag{7.32}$$

Since the traced solution (7.31) must coincide with the original solution (7.23), comparison of coefficients leads to the relations

$$c_1^{(A)} = c_1 \quad ; \quad c_2^{(A)} = c_3 . \tag{7.33}$$

When (7.25,29,32,33) are combined, we obtain the following equations for U_ℓ:

$$1 = [i,-i]_1^{(-)}[-i,i]_1^{(+)} + [i,-i]_1^{(-)}[-i,i]_2^{(+)}U_1U_8 \tag{7.34a}$$

$$0 = [-i,i]_2^{(-)}[-i,i]_1^{(+)}U_3 + [-i,i]_2^{(-)}[i,-i]_2^{(+)}U_8(1 + U_1U_3) \tag{7.34b}$$

$$0 = [i,-i]_1^{(-)}[-i,i]_1^{(+)}U_6 + [i,-i]_1^{(-)}[-i,i]_2^{(+)}U_1(1 + U_6U_8) \tag{7.34c}$$

$$1 = [-i,i]_2^{(-)}[-i,i]_1^{(+)}U_3U_6 + [-i,i]_2^{(-)}[i,-i]_2^{(+)}(1 + U_1U_3)(1 + U_6U_8) \tag{7.34d}$$

The products of the dominancy-changing factors have the following values:

$$[i,-i]_1^{(-)}[-i,i]_1^{(+)} = \exp(-2s)$$

$$[\pm1,\mp i]_1^{(-)}[\mp i,\pm i]_2^{(+)} = \exp(\mp\pi i) = -1$$

$$[-i,i]_2^{(+)}[i,-i]_2^{(-)} = \exp(2s) \quad , \tag{7.35}$$

where

$$2s = -i \oint dt\, T(t)(t^2 + 1)^{1/2} \quad , \tag{7.36}$$

the contour being the path encircling the line $(i,-i)$ in Fig.7.2. This contour integral may be evaluated as follows:

$$-2s = i \oint dt\, T(t)(t + 1/2t + \ldots)$$

$$= \frac{1}{2} i \oint dt\, T(t)t^{-1} + \ldots$$

$$\approx -\pi T(0) = -\pi \left(\frac{\lambda W_{12}}{\Lambda_1 + \Lambda_2} \frac{dr}{dt} \right)_{t=0} \quad , \tag{7.37}$$

where $|t| > 1$ can be chosen for the entire path. The parameter s is the well-known Landau-Zener-Stückelberg (LZS) factor. It is real if Λ_1 and Λ_2 are real at the curve-crossing point. Although the last line in (7.37) is consistent with the tracing procedure applied to obtain the local behavior of the solutions in the neighborhood of $t = \pm i$, it is approximate. One may simply evaluate the integral in (7.37) by a numerical method for better accuracy.

On substituting (7.35) into (7.34), we can solve the latter for U_ℓ within a sign:

$$U_\ell = i[1 - \exp(-2s)]^{1/2} \quad , \quad (\ell = 1,3,6,8) \quad . \tag{7.38}$$

These are the Stokes constants associated with the Stokes lines 1,3,6 and 8.

Since we now know the Stokes constants, we are able to express the solution at D more explicitly. When these Stokes constants are used in the expression for v on the real axis,

$$v = \{C_1 + i[1 - \exp(-2s)]^{1/2}C_3\}(-i,t)_{i1} + \{i[1 - \exp(-2s)]^{1/2}C_1$$
$$+ \exp(-2s)C_3\}(t,-i)_{i2}$$

$$= \{\exp(-s)B_1^{(+)} + [1 - \exp(-2s)]^{1/2}B_2^{(+)}\}(0,t)_{i1} + \{-[1 - \exp(-2s)]^{1/2}B_1^{(+)}$$
$$+ \exp(-s)B_2^{(+)}\}(t,0)_{i2} \quad . \tag{7.39}$$

The second line results when (7.19) is substituted into the first and the square roots of dominancy-changing factors are calculated; see (7.35).

The other branch of solution v_i^* may be traced similarly. We obtain v_i^* at D and on the real axis in the form

$$v^* = \{\exp(-s)B_1^{(-)} + [1 - \exp(-2s)]^{1/2}B_2^{(-)}\}(t,0)_{i1}$$
$$+ \{-[1-\exp(-2s)]^{1/2}B_1^{(-)} + \exp(-s)B_2^{(-)}\}(0,t)_{i2} \quad . \tag{7.40}$$

Finally, by subsituting (7.39,40) into (7.9), we obtain u_i as

$$u_i = e_{i1}[D_1(r_1,r)_1 + D_2(r,r_1)_1]$$
$$+ e_{i2}[D_3(r_2,r)_2 + D_4(r,r_2)_2] \quad , \tag{7.41}$$

where

$$D_1 = (r_c,r_1)_1\{\exp(-s)B_1^{(+)} + [1 - \exp(-2s)]^{1/2}B_2^{(+)}\}$$

$$D_2 = (r_1,r_c)_1\{\exp(-s)B_1^{(-)} + [1 - \exp(-2s)]^{1/2}B_2^{(+)}\} \tag{7.41a}$$

$$D_3 = (r_c,r_2)_2\{-[1-\exp(-2s)]^{1/2}B_1^{(+)} + \exp(-s)B_1^{(-)}\}$$

$$D_4 = (r_2,r_c)_2\{-[1 - \exp(-2s)]^{1/2}B_1^{(-)} + \exp(-s)B_2^{(-)}\} \quad .$$

Equations (7.41,41a) constitute the end result of the tracing across the crossing point.

The scattering-matrix elements can be constructed from (7.41) by imposing the boundary conditions (5.87). Since the general procedure has already been described in Chap.5, only the result is presented below:

$$S_{11} = \exp(2in_1)\{\exp(-2s) + [1 - \exp(-2s)]\exp(-2i\tau)\}$$

$$S_{12} = 2i^{\ell_1 - \ell_2 + 1} \exp[i(\eta_1 + \eta_2)]\exp(-s)[1 - \exp(-2s)]^{1/2} \sin\tau \quad , \qquad (7.42)$$

where

$$\tau = \lambda \int_{r_1}^{r_c} \Lambda_1 \; dx - \lambda \int_{r_2}^{r_c} \Lambda_2 \; dx \quad .$$

The LZS scattering-matrix elements provide qualitatively correct numerical values for transition probabilities [7.3]. There have been some efforts to analyze [7.4,5] and improve [7.6,7] the method. *Child* [7.6] and *Kotova* [7.7] found that there is a small correction factor χ to the phase τ which depends on s as follows:

$$\chi = \arg\{\Gamma(i\delta)\} + \delta - \delta \ln\delta + \pi/4 \quad ,$$

where $\delta = s/\pi$. This factor is generally less than $\pi/4$ in absolute magnitude, but sometimes significantly improves the numerical accuracy of the transition probability as was shown by *Nikitin* and *Reznikov* [7.8] in the case of the $He^+ + Ne(2p^6) \rightarrow He^+ + Ne(2p^5 3s)$ inelastic transition [7.9].

7.2 Applications of the Uniform Semiclassical Theories

In this and following sections we shall apply the semiclassical theories developed in Chap.5 to a two-state curve-crossing problem. They are studied here since there are many practical systems that can be described by a two-state curve crossing model, but the theories apply to noncurve-crossing problems as well.

As was shown in Chap.5, one needs to solve the equation of motion, and particularly (5.53) in the present case. We have already indicated some ways to obtain either approximate or numerical solutions. Here we shall consider approximate methods in more detail. It is useful to define some symbols related to the integral of the matrix $\mathbf{W}$ that will help us simplify equations:

$$I(Z_i, Z_j)_v = (1/2) \int_{v_t}^{v} dv'(Z_i, Z_j)(v') \quad , \qquad (7.43)$$

where

$$Z_j = A_j, B_j \quad , \quad (i,j = 1,2)$$

and hence (Z_i, Z_j) is defined by (5.49). When the upper limit of the integral is $v = \pi/2$, we shall simply denote the integral by $I(Z_j, Z_j)$ without the subscript v.

As was shown before, the integral F_1 of $\frac{1}{2}W$ is also involutional. Let

$$F_1 = \frac{1}{2} \int_{v_t}^{v} dv' \; W(v') \quad . \tag{7.44}$$

Then

$$F_1^2(v) = -f_1(v)I \tag{7.45}$$

with

$$f_1(v) = I(A_1,A_2)_v I(B_1,B_2)_v - I(A_1,B_2)_v I(B_1,A_2)_v \quad . \tag{7.46}$$

The property (7.45) plays an important role in the approximation methods developed subsequently.

7.2.1 The First-Order Magnus Approximation

The first-order Magnus approximation for the equation of motion is

$$X(v) = \exp(F_1)X_0 \quad . \tag{7.47}$$

The lower end v_t of the integral for F_1 may be chosen with the outermost classical turning point. This approximation is simple, but produces reasonable results for transition probabilities generally better than the formulas in (7.42) predict.

To calculate the scattering matrix elements for the problem, it is helpful to digress a little to consider some features of the matrix F_1. Let us construct from (7.45) matrices $X_\pm$ such that

$$X_\pm = F_1 \pm i f_1^{1/2} I \quad , \tag{7.48}$$

which obviously implies that

$$I = (X_+ - X_-)\left[2if_1^{1/2}(v)\right]^{-1} \quad . \tag{7.49}$$

A simple calculation shows that

$$F_1 X_\pm = \pm i f_1^{1/2} X_\pm \quad , \tag{7.50}$$

suggesting that $X_\pm$ may be formally looked upon as the eigenvectors of the matrix F_1 with the eigenvalues $\pm i f_1^{1/2}$. These three equations may be used for calculating the exponential matrix in (7.47):

$$\exp(F_1) = \exp(F_1)(X_+ - X_-)\left[2if_1^{1/2}(v)\right]^{-1}$$

$$= \left\{\exp\left[if_1^{1/2}(v)\right]X_+ - \exp\left[-if_1^{1/2}(v)\right]X_-\right\}\left[2if_1^{1/2}(v)\right]^{-1}$$

$$= \cos f_1^{1/2}(v)I + \left[\sin f_1^{1/2}(v)/f_1^{1/2}(v)\right]F_1(v) \quad .$$

This identity is useful for calculating the scattering-matrix elements following the procedure described in Chap.5. They are as follows:

$$S_{11} = \exp(2in_1) \frac{1 - R_{12}R_{21} \exp[i(\vartheta_{12} - \vartheta_{21})]}{1 + R_{12}R_{21} \exp[i(\vartheta_{12} + \vartheta_{21})]} \tag{7.51a}$$

$$S_{12} = 2i^{\ell_1 - \ell_2 + 1} \exp[i(n_1 + n_2)] \frac{f_1^{-1/2} \tan f_1^{1/2} I(A_1, A_2)}{1 + R_{12}R_{21} \exp[i(\vartheta_{12} + \vartheta_{21})]} \quad , \tag{7.51b}$$

where f_1 means $f_1(v)$ evaluated at $v = \pi/2$, and other symbols are defined by

$$R_{21} = f_1^{-1/2} \tan f_1^{1/2} \left[I^2(A_2, B_1) + I^2(A_2, A_1)\right]^{1/2}$$

$$R_{12} = f_1^{-1/2} \tan f_1^{1/2} \left[I^2(A_1, B_2) + I^2(A_1, B_2)\right]^{1/2} \tag{7.52}$$

$$\vartheta_{21} = \tan^{-1}[I(A_2, B_1)/I(A_1, A_2)]$$

$$\vartheta_{12} = -\tan^{-1}[I(A_1, B_2)/I(A_1, A_2)] \quad . \tag{7.53}$$

The integrals appearing in (7.52,53) may be calculated numerically [7.10]. The S matrix thus calculated generally has better numerical accuracy than the LZS theory S matrix without the phase correction. It must be pointed out that (7.51) holds even for noncurve-crossing situations, whereas the LZS formulas hold only for curve crossing. The accuracy of (7.51) is assessed in comparison with other S matrix formulas later in this chapter.

It must be noted that the S matrix elements obtained above do not apply to cases where the energy is less than the curve-crossing energy, since there is more than one turning point involved in such cases. See Chaps.4,9 for discussions on multiturning-point problems.

7.2.2 The Stationary-Phase Method

The calculation of integrals (7.43) can be performed quite effectively without losing accuracy in curve-crossing cases, if the stationary-phase method is used [5.25,7.11]. Since this idea finds applications in other approximation methods to be discussed later, we shall treat it here. It also yields the S matrix elements in rather transparent forms reminiscent of those in the Landau-Zener-Stückelberg theory.

Let us return to (7.43) and write it out. For example,

$$I(A_1,A_2) = \frac{1}{2} \int_{v_t}^{\infty} du \; \frac{(A_1,A_2)(u)}{1 + u^2}$$

$$= \int_{r_t}^{\infty} dx(du/dx) \; \frac{2w_{12}^2}{[E_1(x) - E_2(x)]^2 + 4w_{12}^2} \; (A_1,A_2)(x) \quad , \qquad (7.54)$$

where we have reverted back to the variable u from the variable v — see (5.53) — and other symbols are defined in Chap.5. The integrand consists of two distinct factors; one is the Wronskian which oscillates and the other is the bell-shaped function which is maximum at a point r_c where $E_1(r) - E_2(r)$ is minimum. As a matter of fact, if a curve-crossing point r_c exists at which

$$E_1(r) - E_2(r) = 0 \quad ,$$

then the factor is maximum at $r = r_c$. Moreover, the oscillatory Wronskian has a stationary phase at the crossing point. This means that the major contribution to the integral comes from the vicinity of the crossing point. It is easily seen that other integrals behave similarly. In view of this feature, we may use the method of stationary phase for evaluating integrals.

First, we write the Wronskians in the following forms by expressing the functions A_j and B_j in terms of amplitudes and phases [2.14,7.11]:

$$(A_1,A_2) = M_1 \cos\theta_1 \cos\theta_2 - M_2 \cos\theta_1 \cos\bar{\psi}_2 + M_3 \cos\bar{\psi}_1 \cos\theta_2$$

$$(A_1,B_2) = M_1 \cos\theta_1 \sin\theta_2 - M_2 \cos\theta_1 \sin\bar{\psi}_2 + M_3 \cos\bar{\psi}_1 \sin\theta_2$$

$$(B_1,A_2) = M_1 \sin\theta_1 \cos\theta_2 - M_2 \sin\theta_1 \cos\bar{\psi}_2 + M_3 \sin\bar{\psi}_1 \sin\theta_2$$

$$(B_1,B_2) = M_1 \sin\theta_1 \sin\theta_2 - M_2 \sin\theta_1 \sin\bar{\psi}_2 + M_3 \sin\bar{\psi}_1 \sin\theta_2 \qquad (7.55)$$

where with $x_j = \lambda^{3/2} t_j$

$$M_1 = (\pi/2)[(x_1''/x_1') - (x_2''/x_2')](x_1' x_2')^{1/2} a_1(x_1) a_2(x_2)$$

$$M_2 = \pi(x_2'/x_1')^{1/2} a_1(x_1) b_2(x_2)$$

$$M_3 = \pi(x_1'/x_2')^{1/2} a_2(x_2) b_1(x_1)$$

$$a_j(x_j) = [Ai^2(-x_j) + Bi^2(-x_j)]^{1/2}$$

$$b_j(x_j) = [Ai'^2(-x_j) + Bi'^2(-x_j)]^{1/2}$$

$$\theta_j = \tan^{-1}[Bi(-x_j)/Ai(-x_j)]$$

$$\bar{\psi}_j = \tan^{-1}[Bi'(-x_j)/Ai'(-x_j)] \quad . \tag{7.56}$$

Here the primes mean differentiations.

Now if we define new phases,

$$\Delta^{\pm} = \theta_2 \pm \overset{\bullet}{\theta}_1$$

$$\phi_i = \bar{\psi}_i - \theta_i \quad , \quad (i = 1,2) \quad , \tag{7.57}$$

(7.55) can be reduced to much simpler forms as follows:

$$(A_1, B_2) = p^- \sin(\Delta^- + \xi^-) + p^+ \sin(\Delta^+ + \xi^+)$$

$$(A_1, B_2) = -p^- \cos(\Delta^- + \xi^-) - p^+ \cos(\Delta^+ + \xi^+)$$

$$(B_1, A_2) = p^- \cos(\Delta^- + \xi^-) - p^+ \cos(\Delta^+ + \xi^+)$$

$$(B_1, B_2) = p^- \sin(\Delta^- + \xi^-) - p^+ \sin(\Delta^+ + \xi^+) \quad , \tag{7.58}$$

where

$$p^{\pm} = (g^{\pm 2} + g_1^2)$$

$$\xi^{\pm} = \tan^{-1}(g_1/g^{\pm})^{1/2}$$

$$g_1 = \frac{1}{2}(M_1 - M_2 \cos\phi_2 + M_3 \cos\phi_1)$$

$$g^{\pm} = \frac{1}{2}(M_2 \sin\phi_2 \mp M_3 \sin\phi_1) \quad . \tag{7.59a}$$

Note that no approximations have been made so far for the functions A_j and B_j.

Now, to estimate the terms involved in (7.56), we use the asymptotic formulas for phases and amplitudes [2.14]:

$$\theta_j(x) \underset{x \to \infty}{\sim} \frac{\pi}{4} - \frac{2}{3}x^{3/2}\left[1 - \frac{5}{4}(2x)^{-3} + \ldots\right]$$

$$a_j^2(x) \underset{x \to \infty}{\sim} (1/\pi x^{1/2})\left[1 - \frac{5}{4}(2x)^{-3} + \ldots\right]$$

$$b_j^2(x) \underset{x \to \infty}{\sim} (x^{1/2}/\pi)\left[1 + \frac{7}{4}(2x)^{-3} + \dots\right] \qquad (7.59b)$$

and the relation between phases

$$\phi_i = \bar{\psi}_i - \theta_i = \sin^{-1}(\pi a_i b_i)^{-1} \quad .$$

By using these formulas, we obtain the following relations in the limit of $\lambda \to \infty$:

$$\Delta^{\pm}(u) \simeq \frac{2}{3}(x_1^{3/2} \pm x_2^{3/2})$$

$$= \lambda \left[\int_{r_1}^{r} \Lambda_1 dr' \pm \int_{r_2}^{r} \Lambda_2 dr'\right] \quad , \qquad (7.60)$$

$$\phi_i \simeq \pi/2 \quad .$$

In the same limit g_1 and g^* may be written as

$$g_1 \simeq \frac{1}{2} M_1$$

$$\approx \frac{1}{4}\left[(x_1''/x_1') - (x_2''/x_2')\right](\lambda^2 \Lambda_1 \Lambda_2)^{-1/2} \qquad (7.61)$$

$$g^{\pm} \simeq \frac{1}{2}(M_2 \mp M_3)$$

$$\approx \frac{1}{2}(\Lambda_2 \mp \Lambda_1)(\Lambda_1 \Lambda_2)^{-1/2} \quad . \qquad (7.62)$$

Since the difference between derivatives of $\ln(x_i')$, $(i = 1,2)$, is much smaller than $(\lambda^2 \Lambda_1 \Lambda_2)^{1/2}$, g_1 is negligible compared with g^+ and g^-. Then $\xi^{\pm}$ becomes negligible, too. Therefore in this approximation the phases $\xi^{\pm}$ may be neglected and the amplitudes $p^{\pm}$ may be further approximated by

$$p^{\pm} \simeq g^{\pm} \quad .$$

Now let us compare the amplitudes $g^{\pm}$ and phase $\Delta^{\pm}$ respectively. Since $\Lambda_1(x_c) \simeq \Lambda_2(x_c)$ at $x = x_c$, it is clear from (7.61,62) that

$$g^+(0) \approx 0$$

$$g^-(0) \approx 1 \quad ,$$

and from (7.60) that at the curve-crossing point

$$|\Delta^-(0)| < |\Delta^+(0)| \quad .$$

Moreover, the term with phase $\Delta^{+}(u)$ oscillates more rapidly than that with phase $\Delta^{-}(u)$. Therefore, the former will produce destructive interferences more effectively for the integral than the latter. Consequently the term with $\Delta^{+}(u)$ may be neglected in the Wronskians. Thus we have

$$(A_1, A_2) \approx \sin\Delta^{-}(u) \quad ,$$

$$(A_1, B_2) \approx -\cos\Delta^{-}(u) \quad ,$$

$$(B_1, A_2) \approx \cos\Delta^{-}(u) \quad ,$$

$$(B_1, B_2) \approx \sin\Delta^{-}(u) \quad . \tag{7.63}$$

When these formulas are put into (7.43), the stationary-phase approximation formulas follow for the integrals. They are special cases of the following integrals:

$$c_i(u) = \frac{1}{2} \int_{u_t}^{u} du \, h_i(u)(1 + u^2)^{-1} \cos\psi(u) \tag{7.64a}$$

$$s_i(u) = \frac{1}{2} \int_{u_t}^{u} du \, h_i(u)(1 + u^2)^{-1} \sin\psi(u) \quad , \quad (i = 1,2) \quad , \tag{7.64b}$$

where

$$h_1(u) = 1 \quad ; \quad h_2(u) = (1 + u^2)^{-1/2} \quad ; \quad h_3(u) = u(1 + u^2)^{-1/2} \quad ;$$

$$\psi(u) = \Delta^{-}(u) - \Delta^{-}(0) \quad .$$

As a matter of fact, for the purpose of this subsection we need only the case of $h_1(u) = 1$ in the integrals, but the integrals are given in the general forms, since they will appear later when more elaborate approximations are developed.

In the approximation made for the Wronskians the matrices $\boldsymbol{P}_{12}$ and $\boldsymbol{P}_{21}$ take especially interesting forms which remind us of orthogonal transformations in two-dimensional space:

$$\boldsymbol{P}_{21} \approx - \begin{pmatrix} \cos\Delta^{-}(u) & \sin\Delta^{-}(u) \\ -\sin\Delta^{-}(u) & \cos\Delta^{-}(u) \end{pmatrix} \equiv -R(u) \tag{7.65a}$$

$$\boldsymbol{P}_{12} \approx - \begin{pmatrix} \cos\Delta^{-}(u) & -\sin\Delta^{-}(u) \\ \sin\Delta^{-}(u) & \cos\Delta^{-}(u) \end{pmatrix} \equiv -\tilde{R}(u) \quad . \tag{7.65b}$$

Therefore in the present stationary-phase approximation the matrix $W(u)$ is given by

$$W(u) \approx \begin{pmatrix} 0 & R(u) \\ -\tilde{R}(u) & 0 \end{pmatrix} \equiv D[\Delta^-(u)] \quad . \tag{7.65c}$$

It is then easy to see that

$$W^2(u) = -I \quad ,$$

which shows that the approximation preserves the involutionality of the matrix. When the matrix is integrated over u, the resulting matrix F_1 has the same mathematical structure as $D(u)$ except for the different argument of D:

$$F_1 = \lambda_1 D[\gamma_1 + \Delta_0] \quad , \tag{7.66}$$

where

$$\Delta_0 = \Delta^-(0)$$

$$\lambda_1 = [c_1^2(\infty) + s_1^2(\infty)]^{1/2}$$

$$\gamma_1 = \tan^{-1}[s_1(\infty)/c_1(\infty)] \quad . \tag{7.67a}$$

Note that the argument of D has changed from $\Delta^-(u)$ to a new argument $\gamma_1 + \Delta_0$ when the integration is performed. The involutionality of the matrix F_1 is also preserved.

The f_1 defined by (7.46) is now given in terms of integrals $c_1(u)$ and $s_1(u)$,

$$f_1(u) = c_1^2(u) + s_1^2(u) = \lambda_1^2(u) \quad , \tag{7.67b}$$

and R_{12}, R_{21}, ϑ_{12}, and ϑ_{21} are easily found to take the forms

$$R_{12} = R_{21} = \tan\lambda_1 \quad , \qquad [\lambda_1 \equiv \lambda_1(\infty)] \quad ,$$

$$\vartheta_{21} = -\vartheta_{12} = \Delta^-(\infty) + \frac{\pi}{2} \equiv \vartheta \quad .$$

When these results are used in (7.51), we find the S-matrix elements in the following rather simple forms:

$$S_{11} = \exp(2i\eta_1) \; [\cos^2\lambda_1 - \sin^2\lambda_1 \exp(-2i\vartheta)] \tag{7.68a}$$

$$S_{12} = i^{\ell_1 - \ell_2 + 1} \exp[i(\eta_1 + \eta_2)]\sin(2\lambda_1)\sin\tau_c \quad , \tag{7.68b}$$

where

$$\tau_c = \gamma_1 + \Delta_0 \quad .$$

The S matrix elements obtained above correspond to those in (7.42) of the LZS theory and are obviously as simple as the latter, yet in fact they are also more accurate according to a numerical test [5.25,7.11]. It should be noted that Δ_0 is simply the Stückelberg phase τ in (7.42) and hence γ_1 is a correction factor to the Stückelberg phase. It is less than $\pi/2$ and tends to zero as the coupling constant vanishes.

7.2.3 A Resummation Method

The first-order Magnus approximation gives reasonably accurate S matrix elements, but there is still room to improve their accuracy. A considerable improvement in accuracy can be achieved if a higher-order Magnus approximation is employed. However, since it involves an infinite series, some ways to resum the series are required. Fortunately, the involutionality of the matrices involved comes to rescue us in the present case. The reader familiar with Sect.4.2 can probably see right away how it can be done. The method [7.10] is completely parallel to that presented in Sect.4.2 for eleastic scattering. We set

$$X(v) = \exp[F(v)]X_0 \quad , \tag{7.69}$$

where

$$X_0 = \{\alpha_1^0, \ 0, \ \alpha_2^0, \ 0\}$$

is the initial condition. Substitution of (7.69) into the equation of motion (5.53b) yields [cf. (4.39)],

$$\frac{d}{dv} F = \{F(\exp F - 1)^{-1} , \ \tfrac{1}{2}\omega\}$$

$$= \tfrac{1}{2}\omega - \tfrac{1}{4}[F,\omega] + \sum_{n=1}^{\infty} \frac{(-)^{n-1}B_n}{(2n)!} \{F^{2n}, \tfrac{1}{2}\omega\} \tag{7.70}$$

$$\equiv H(F,\omega) \quad ,$$

where the various notations have been defined in previous chapters. This equation may be solved by iteration. For this purpose we define a sequence $\{F_0, F_1, F_2, \ldots, F_m, \ldots\}$ where $F_0 = 1$ and the other elements are defined by

$$\frac{d}{dv} F_m = H(F_{m-1}, \boldsymbol{\mathcal{W}})$$

$$\equiv H_m \quad . \tag{7.71}$$

We then see that (7.44) is recovered for m = 1, which is the first-order Magnus approximation. The second-order Magnus approximation results for m = 2 and it produces relatively simple formulas, thanks to the involutionality of the matrix F_1. Since the procedure of resummation is the same as in Sect.4.2, we briefly sketch here the basic points with relevant equations.

First define the Lie product,

$$G_n = \{F_1^{2n}, \boldsymbol{\mathcal{W}}\} \quad (n = 1,2,\dots) \tag{7.72}$$

which may be written in the following recursion relation:

$$G_n = \{F_1^2, G_{n-1}\}$$

$$= [F_1, [F_1, G_{n-1}]]$$

$$= -2f_1 G_{n-1} - 2F_1 G_{n-1} F_1$$

for which (7.45) has been used. Substitute

$$G_{n-1} = -2f_1 G_{n-2} - 2F_1 G_{n-2} F_1$$

into the last line of the equation above and then make use of (7.45) again. Then there follows

$$G_n = -4f_1 G_{n-1} \quad ,$$

which can be easily solved,

$$G_n = (-4f_1)^{n-1} G_1$$

$$= 2(-4f_1)^{n-1} F_1 [F_1, \boldsymbol{\mathcal{W}}] \quad . \tag{7.73}$$

This solution of the recursion relation enables us to resum the infinite series of Lie products appearing in (7.70):

$$\sum_{n=1}^{\infty} \frac{(-)^{n-1} B_n}{(2n)!} \{F_1^{2n}, \boldsymbol{\mathcal{W}}\} = \sum_{n=1}^{\infty} \frac{(-)^{n-1} B_n (-4f_1)^{n-1}}{(2n)!} G_1$$

$$= -G_1 (4f_1)^{-1} \sum_{n=1}^{\infty} \frac{(-)^{n-1} B_n (2if_1^{1/2})^{2n}}{(2n)!}$$

$$= G_1(4f_1)^{-1}(1 - f_1^{1/2} \cot f_1^{1/2}) \quad ,$$

where we have made use of the well-known relation [2.2]

$$\frac{t}{\exp(t) - 1} = (t/2i)\cot(t/2i) - t/2 = 1 - t/2 + \sum_{n=1}^{\infty} \frac{(-)^{n-1}B_n}{(2n)!} t^{2n} \quad .$$

Therefore we finally obtain the second approximation

$$F_2(v) = \int_{v_t}^{v} dv \, H_2(F_1, \mathcal{W}) \quad , \tag{7.74a}$$

where the explicit form for H_2 is now

$$H_2 = \tfrac{1}{2}\mathcal{W} - \tfrac{1}{4}[F_1, \mathcal{W}] + (4f_1)^{-1}(1 - f_1^{1/2} \cot f^{1/2}) F_1[F_1, \mathcal{W}] \quad . \tag{7.74b}$$

Unlike the second approximation for the equation of motion (4.39) for elastic scattering, the second F_2 is no longer involutional and hence a similar calculation cannot be made for the third or higher approximations. However, it is found empirically that the second approximation is already sufficiently accurate so that a higher-order approximation is not necessary [7.11].

The S matrix elements for the second-order approximation can be calculated from (5.89) with a_{ij} determined by

$$a_{i1} = \left\{ \exp\left[\int_{v_t}^{\pi/2} dv \, H_2(v) \right] \right\}_{i1}$$
$$\qquad\qquad\qquad (i = 1,2,3,4) \tag{7.75}$$
$$a_{i2} = \left\{ \exp\left[\int_{v_t}^{\pi/2} dv \, H_2(v) \right] \right\}_{i3} \quad .$$

The lower end v_t of the integrals appearing above is chosen the same as before with the value of v corresponding to the largest turning point for the potentials involved.

The results up to now are valid regardless of whether a curve-crossing point exists or not. However, if there is a curve-crossing point, further approximations may be made without losing accuracy as in the case of the first-order Magnus approximation. Since the approximations yield S matrix elements quite simple to compute, but generally superior to the Landau-Zener-Stückelberg formula, we discuss them here. These approximations are given by the procedure described in Sect.7.2.2, where the asymptotic formu-

las are used for Ai(-t) and Bi(-t) and the terms involving the stationary
phases are retained for calculating various integrals.

Thus, by following the method described there and substituting (7.63,64)
into (7.74), we find that F_2 consists of the following three integrals in
addition to c_1 and s_1:

$$q_1(v) = \frac{1}{2} \int_{v_t}^{v} dv' \ f_1(v')\sin\phi(v') \tag{7.76a}$$

$$q_2(v) = \frac{1}{2} \int_{v_t}^{v} dv'(1 - f_1^{1/2} \cot f_1^{1/2})s_1(v')\sin\phi(v') \tag{7.76b}$$

$$q_3(v) = \frac{1}{2} \int_{v_t}^{v} dv'(1 - f_1^{1/2} \cot f_1^{1/2})c_1(v')\sin\phi(v') \ , \tag{7.76c}$$

where

$$\phi(v) = \gamma_1(v) - \bar{\Delta}(v) \ .$$

Thus, after some calculations it is found that the second approximation F_2
has the form

$$F_2(v) = f_1(v)D[\gamma_1(v) + \Delta_0] + \bar{F}_2(v) \tag{7.77}$$

$$\bar{F}_2(v) = \begin{pmatrix} 0 & q_1(v) & -q_2(v) & q_3(v) \\ -q_1(v) & 0 & -q_3(v) & -q_2(v) \\ q_2(v) & q_3(v) & 0 & q_1(v) \\ -q_3(v) & q_2(v) & -q_1(v) & 0 \end{pmatrix} \ , \tag{7.78}$$

Note that the first term on the right-hand side of (7.77) is the first-order
Magnus approximation and in fact is the same as (7.66), and therefore the
second term consitutes the correction to the first-order Magnus approximation.

The S matrix elements can be obtained from the matrices given above by
putting $v = \pi/2$ and calculating

$$a_{ij} = \{\exp[F_2(\pi/2)]\}_{ij} \ .$$

7.2.4 Modified Magnus Approximation

Although the resummation method described in Sect.7.2.3 is sufficiently accurate, it is not the only possible way to solve a curve-crossing problem.
In fact, if one wants to apply the method to a situation where more than

one curve cross each other, it is not the most convenient. Here we describe a method [7.10] as accurate as, and more convenient than, the resummation method, which can be used as the foundation on which to build an approximate multichannel theory. This method also exploits the involutionality of W which helps in obtaining very compact, often analytic, formulas.

We have pointed out that the first-order Magnus approximation requires some improvement in accuracy. We approach this improvement of the solution in the following way. First we note that the usual Magnus approximation

$$X(v) = \exp(F_1)X_0$$

is a partial resummation of the iterative (perturbation) series generated from the integral equation equivalent to (5.53b),

$$X(v) = X_0 + \int_{v_t}^{v} dv' \tfrac{1}{2}W(v')X(v') \quad .$$

This is a Volterra-type integral equation and therefore its iterative (Neumann) series will always converge if the appropriately defined norm of the kernel is less than 1:

$$\| \tfrac{1}{2} W \| \leq 1 \quad .$$

However, we have the relation

$$W^2(v) = -I \quad ,$$

which suggests that the norm of the kernel may be larger than 1 (in fact, $\| \tfrac{1}{2} W \| \leq \pi/2$). Therefore the iterative series may not be very accurately summed into such a simple exponential form and any partial resummation thereof may not be so reliable. However, we also note that the involutionality holds true for any v and in particular at v = 0, the curve-crossing point, i.e.,

$$W_0^2 \equiv W^2(0) = -I \quad .$$

In view of these two relations, we may expect that

$$\| \tfrac{1}{2}\{W(v) - W(0)\} \| < 1 \quad . \tag{7.79}$$

To exploit this point, we write

$$X(v) = \exp[\tfrac{1}{2}W_0(v - v_t)]Y(v) \quad . \tag{7.80}$$

On substitution of (7.80) into (5.53b), we obtain the differential equation for Y(v):

$$\frac{d}{dv} Y(v) = \omega^{(1)}(v)Y(v) \quad , \tag{7.81}$$

where

$$\omega^{(1)}(v) = \exp[-\tfrac{1}{2}\omega_0(v - v_t)] \tfrac{1}{2} [\omega(v) - \omega_0]\exp[\tfrac{1}{2}\omega_0(v - v_t)] \quad . \tag{7.82}$$

Here we note that owing to the involutionality of ω

$$\exp[\pm\tfrac{1}{2}\omega_0(v - v_t)] = \cos[(v - v_t)/2]\mathbf{I} \pm \sin[(v - v_t)/2]\omega_0 \quad . \tag{7.83}$$

In view of (7.79) the iterative series generated from the integral equation

$$Y(v) = X_0 + \int_{v_t}^{v} dv' \; \omega^{(1)}(v')Y(v')$$

is expected to be fast convergent, and the first-order Magnus approximation to (7.81) to be of better quality.

Based on this intuitive argument, we then obtain the approximate solution in the first-order Magnus approximation,

$$X(v) = \exp[\tfrac{1}{2}\omega_0(v - v_t)]\exp[\mathbf{P}(v)]X_0 \quad , \tag{7.84}$$

where

$$\mathbf{P}(v) = \int_{v_t}^{v} dv' \; \omega^{(1)}(v')$$

$$= \tfrac{1}{2} \int_{v_t}^{v} dv'\left\{[\omega(v') - \omega_0] + \tfrac{1}{2} \sin(v' - v_t)[\omega,\omega_0] \right. \tag{7.85}$$

$$\left. + \sin^2 \tfrac{1}{2}(v' - v_t)\omega_0[\omega_0,\omega]\right\} \quad .$$

The structure of (7.85) is somewhat reminiscent of (7.74a,b). The relation between (7.85) and (7.44) for the usual Magnus approximation can be seen if we use the Baker-Hausdorff formula [4.29,7.12] for (7.84):

$$X(v) = \exp\left\{\tfrac{1}{2}\omega_0(v - v_t) + \mathbf{P}(v) + \tfrac{1}{2} [\tfrac{1}{2}\omega_0(v - v_t) , \mathbf{P}(v)] + \ldots\right\}X_0$$

$$= \exp(\mathbf{F}_1 + \Delta\mathbf{F})X_0 \quad ,$$

where

$$\Delta\mathbf{F} = \tfrac{1}{4} \int_{v_t}^{v} dv' \; \sin(v' - v_t)[\omega,\omega_0] + \tfrac{1}{2} \int_{v_t}^{v} dv' \; \sin^2 \tfrac{1}{2} (v - v_t)\omega_0[\omega_0,\omega]$$

$$+ \tfrac{1}{4} (v - v_t)[\omega_0,\mathbf{P}] + \ldots \quad .$$

Therefore ΔF constitutes a correction to F_1. We are now ready to calculate a_{ij} for the S matrix elements. It is more convenient to use (7.84) with the first exponential term replaced by (7.83),

$$X(v) = \left[\cos \frac{1}{2}(v - v_t) I + \sin \frac{1}{2}(v - v_t) W_0\right]\exp[P(v)]X_0 \quad , \tag{7.86}$$

where the matrix $P(v)$ is computed by means of (7.85). We compute a_{ij} with the formula

$$a_{ij} = \left\{[\cos(\pi/4 - v_t/2) I + \sin(\pi/4 - v_t/2) W_0]\exp[P(\pi/2)]\right\}_{i,2j-1} \quad ,$$
$$(i = 1,2,3,4; \quad j = 1,2) \quad . \tag{7.87}$$

Calculating a_{ij} now requires integrals of W itself or its products with sinusoidal functions and hence is no more difficult than the integrals of the first-order Magnus approximation. The S matrix calculated from (7.87) is now much improved in accuracy.

It is possible to compute the matrix P by using the stationary-phase method described in Sect.7.2.2 without losing numerical accuracy. When this method is applied to calculate P, various Wronskians are replaced with their approximate formulas (7.63). Then we obtain W_0 in the form

$$W_0 = W(0) = D[\Delta_0] \quad , \tag{7.88}$$

as is obvious from the previous section. Note that in the same approximation

$$W(u) = D[\Delta^-(u)] \quad .$$

Then with the definitions

$$\lambda_2 = \left\{\left[\frac{1}{2}(\pi/2 - v_t) - c_1(\infty)\right]^2 + s_2^2(\infty) + s_3^2(\infty)\right\}^{1/2} \tag{7.89}$$

$$\zeta = \sin^{-1}[(s_2^2(\infty) + s_3^2(\infty))^{1/2}/\lambda_2] \tag{7.90}$$

$$\alpha = \tan^{-1}[s_3(\infty)/s_2(\infty)] \tag{7.91}$$

a much simplified form for P follows:

$$P(\pi/2) = -\lambda_2\{\cos\zeta \ D[\Delta_0] + \sin\zeta \ \cos(\pi/2 + \alpha + v_t)D[\pi/2 + \Delta_0]$$

$$- \sin\zeta \ \sin(\pi/2 + \alpha + v_t)D[\pi/2 + \Delta_0]D[\Delta_0]\} \quad . \tag{7.92}$$

For (7.92) we have made use of the fact that $P(\pi/2)$ becomes involutional in the present approximation, i.e.,

$$P^2(\pi/2) = -\lambda I \quad ,$$

which implies

$$\exp[P(\pi/2)] = \cos\lambda_2 \, I + \lambda_2^{-1} \sin\lambda_2 \, P(\pi/2) \quad . \tag{7.93}$$

It is now trivial to obtain a_{ij} by substituting (7.92,93) into (7.87). After lengthy mathematical manipulation, we find the S-matrix element S_{12} in a form similar to (7.68b):

$$S_{12} = -i^{\,\ell_1-\ell_2+1} \, \exp[i(n_1 + n_2)] \, \sin(2\lambda_2)f \quad , \tag{7.94}$$

where

$$f = \cos(\pi/2 + v_t) \, \cos\xi \, \sin\Delta_0 + \sin\xi \, \cos\alpha \, \cos\Delta_0$$

$$+ \cot(2\lambda_2) \, \sin(\pi/2 + v_t) \, \sin\Delta_0$$

$$- \tan\lambda_2 \, \sin\xi \, \sin\alpha[\sin\xi \, \cos(\pi/2 + \alpha + v_t)\sin\Delta_0 - \cos\xi \, \cos\Delta_0] \quad , \tag{7.95}$$

$$\xi = \tan^{-1}\{[s_1^2(\infty) + s_2^2(\infty)]^{1/2}/[2^{-1}(\pi/2 - v_t) - c_1]\} \quad . \tag{7.96}$$

The elastic component S_{11} may be similarly calculated. The first-order Magnus approximation formula for S_{12} arises from (7.94-96) if we put

$$s_2(\infty) = s_3(\infty) = 0 \qquad \text{and} \qquad v_t = \pi/2 \quad ,$$

which yield $\xi = -\gamma_1$ and $\alpha = 0$. We must take $v_t = \pi/2$, since the term $\mathscr{W}(0)$ does not appear in the first-order Magnus approximation. If these values are put into the formula for f, then

$$f = \sin(\Delta_0 + \gamma_1)$$

and λ_2 reduces to λ_1. The S_{12} given above is superior to (7.68b).

7.2.5 Numerical Comparisons

To assess the accuracy and hence the utility of the results obtained up to now, we choose a curve-crossing model which describes reasonably well the $2p^6 - 2p^5 3s$ inelastic transition in the He^+-Ne system studied experimentally [7.13]. The potential functions have the following forms [7.9]:

$$V_{11}(r) = a_1 r^{-1} \exp(-r/a_2)$$

$$V_{22}(r) = (a_1 r^{-1} - a_3)\exp(-r/a_2) + 16.8 \text{ eV}$$

$$V_{12}(r) = a_4 \exp(-r/a_5) \quad , \tag{7.97}$$

Table 7.1. Potential parameters in atomic units

a_1	21.1
a_2	0.678
a_3	12.1
a_5	0.667
r_c	2.05

Table 7.2. Comparison of transition probabilities using various methods at E = 70.9 eV and ℓ = 310 in terms of coupling paramter a_4[a]

a_4 [a.u.]	Exact[b]	UWKBNS[c]	UWKBMM[d]	Eq. (7.94)	Eq. (7.68)	LZS[e]
0.01	0.0008	0.0	0.0003	0.0008	0.0002	0.0002
0.1	0.0732	0.0861	0.0832	0.0761	0.0283	0.0211
0.2	0.2704	0.2881	0.2852	0.2783	0.1334	0.0977
0.3	0.5320	0.5439	0.5426	0.5407	0.3280	0.2429
0.4	0.7813	0.7859	0.7881	0.7843	0.5772	0.4337
0.5	0.9494	0.9501	0.9531	0.9435	0.8084	0.6099
0.6	0.9958	0.9948	0.9985	0.9828	0.9458	0.7069
0.7	0.9177	0.9163	0.9207	0.9029	0.9435	0.6948
0.8	0.7458	0.7446	0.7519	0.7342	0.8006	0.5909
0.9	0.5301	0.6292	0.5415	0.5234	0.5597	0.4414
1.0	0.3219	0.3218	0.3375	0.3185	0.2929	0.2927
1.2	0.0551	0.0548	0.0651	0.0517	0.0	0.0926
1.4	0.0006	0.0007	0.0005	0.0026	0.2368	0.0192
1.6	0.0249	0.0250	0.0391	0.0441	0.0784	0.0025
1.8	0.0292	0.0288	0.0642	0.0587	0.0625	0.0002
2.0	0.0127	0.0121	0.0439	0.0349	0.0441	0.0

[a] See Table 7.1 for other potential parameters. [b] Numerical solution of Schrödinger equations. [c] Numerical solution of uniform WKB theory (5.53). [d] The modified Magnus approximation of uniform WKB theory (7.86). [e] The Landau-Zener-Stückelberg theory (7.42).

where parameters a_1, a_2, a_3, and a_5 are summarized in Table 7.1. Various transition probabilities are calculated as a function of the coupling constant a_4. They are summarized in Table 7.2. The entries in Table 7.2 compare the transition probabilities $|S_{12}|^2$ for various values of a_4 obtained by numerical "exact" solution, the uniform WKB numerical solution, the modified Magnus approximation, the stationary-phase method for the first-order Magnus approximation, and the modified Magnus approximation as well as the LZS formula. From the comparison we see that (7.94) works best, and much better than the uncorrected LZS formula.

7.3 The Diabatic Representation Approach

The semiclassical method developed in Sect.5.5 is applied in this section. To keep the presentation uniform, we shall discuss the same two-state problem as in the previous section. In the present approach we need to solve (5.104). The matrix $\mathbf{R}$ or $\mathbf{M}_\infty$ is again involutional and the techniques developed in the previous sections all apply with appropriate changes in notations regarding momenta and related quantities. Here we shall start the discussion with the calculation of the Magnus approximation (5.115). It may be evaluated by numerically computing the quadratures involved. However, since our aim here is not such numerical procedures, but useful approximation methods, we shall sketch a way to evaluate the integrals approximately. The idea used is the same as that for the stationary-phase method. Since we are considering a curve-crossing problem, the major contribution to the integrals in question comes from the vicinity of the curve-crossing point at which the phase difference of the WKB wave functions is stationary.

The asymptotic formulas for A_j and B_j in the present approach are

$$A_j(r) \sim [\pi p_j(r)]^{-1/2} \sin(x_j + \pi/4)$$

$$B_j(r) \sim [\pi p_j(r)]^{-1/2} \cos(x_j + \pi/4) \quad ,$$

where

$$x_j = \lambda \int_{r_j}^{r} dx' \, p_j(x') \quad .$$

Substitution of these asymptotic formulas into (5.103) for R_{12} and R_{21} yields

$$R_{12} = \frac{1}{2} [p_1(r)p_2(r)]^{-1/2}(\mathbf{R}^{(+)} + \mathbf{R}^{(-)})$$

$$R_{21} = \frac{1}{2} [p_1(r)p_2(r)]^{-1/2}(\mathbf{R}^{(+)} - \tilde{\mathbf{R}}^{(-)}) \tag{7.98}$$

where

$$\mathbf{R}^{(+)} = \begin{pmatrix} -\cos\Delta^{(+)}(r) & \sin\Delta^{(+)}(r) \\ \sin\Delta^{(+)}(r) & \cos\Delta^{(+)}(r) \end{pmatrix} \tag{7.99a}$$

$$\mathbf{R}^{(-)} = \begin{pmatrix} -\sin\Delta^{(-)}(r) & -\cos\Delta^{(-)}(r) \\ \cos\Delta^{(-)}(r) & -\sin\Delta^{(-)}(r) \end{pmatrix} \tag{7.99b}$$

and $\tilde{\mathbf{R}}^{(-)}$ is the transpose of $\mathbf{R}^{(-)}$. The phases $\Delta^{(\pm)}(r)$ are defined by

$$\Delta^{(\pm)}(r) = \lambda \int_{r_1}^{r} dx' \ p_1(x') \pm \lambda \int_{r_2}^{r} dx' \ p_2(x') \quad .$$

Note that these phases are different from the phases $\Delta^{\pm}(u)$ in (6.60) in that they are given in terms of diabatic momenta instead of adiabatic momenta.

The phase $\Delta^{(+)}$ has no stationary-phase point, whereas the phase $\Delta^{(-)}$ is stationary at the curve-crossing point r_c, i.e., at $r = r_c$

$$p_1(r_c) = p_2(r_c) \quad .$$

Since the sinusoidal functions with the phase $\Delta^{(+)}$ oscillate rapidly, the contribution from $\mathbf{R}^{(+)}$ is expected to be much smaller than that from $\mathbf{R}^{(-)}$ and hence we may neglect the former in comparison with the latter. Consequently, we may approximate $\mathbf{R}_{12}$ and $\mathbf{R}_{21}$ by

$$\mathbf{R}_{12} = \frac{1}{2} \ [p_1(x)p_2(x)]^{-1/2}\mathbf{R}^{(-)}$$

$$\mathbf{R}_{21} = -\frac{1}{2} \ [p_1(x)p_2(x)]^{-1/2}\tilde{\mathbf{R}}^{(-)} \quad ,$$

which puts the matrix $\mathbf{M}_\infty$ in the form

$$\mathbf{M}_\infty = \frac{1}{2} \ \lambda w_{12}[p_1(r)p_2(r)]^{-1/2} \begin{pmatrix} 0 & -\tilde{\mathbf{R}}^{(-)} \\ \mathbf{R}^{(-)} & 0 \end{pmatrix} \quad . \tag{7.100}$$

It is useful to cast this matrix in a form completely isomorphic to the matrix $D[\Delta^{(-)}(u)]$ in (7.65). Therefore, if we define a new phase μ by

$$\mu(r) = \Delta^{(-)}(r) - \pi/2 \quad ,$$

the matrix $\mathbf{R}^{(-)}$ takes the same form as $-\tilde{\mathbf{R}}(u)$ according to (7.65a,b)

$$\mathbf{R}^{(-)} = - \begin{pmatrix} \cos\mu & -\sin\mu \\ \sin\mu & \cos\mu \end{pmatrix} \quad .$$

This form of the matrix suggests that $\mathbf{M}_\infty$ may be expressed in terms of the matrix $D[\mu(r)]$ as follows:

$$\mathbf{M}_\infty = \frac{1}{2} \ \lambda^2 w_{12}(p_1 p_2)^{-1/2}D[\mu(r)] \quad , \tag{7.101}$$

where $D[\mu(r)]$ is defined by (7.65c).

It is now easy to see that the integral of the matrix M_∞ is composed of the two integrals below:

$$s_0 = (\lambda^2/2) \int_{r_t}^{\infty} dx \, w_{12}(p_1 p_2)^{-1/2} \sin\phi(x)$$

$$c_0 = (\lambda^2/2) \int_{r_t}^{\infty} dx \, w_{12}(p_1 p_2)^{-1/2} \cos\phi(x) \quad , \qquad (7.102)$$

where

$$\phi(r) = \lambda \int_{r_c}^{r} dx \left(p_1(x) - p_2(x) \right) \quad .$$

It is now very easy to find the integral of M_∞ by comparing (7.101) with (7.65c) and recalling (7.66) along with the definitions of the symbols, (7.67a), accompanying it. We find the integral S_1 of M_∞ in the form,

$$S_1 = aD(\tau_0) \qquad (7.103)$$

where

$$a = (s_0^2 + c_0^2)^{1/2} \quad ,$$

$$\mu_0 = \Delta^{(-)}(r_c) - \pi/2 \quad .$$

$$\tau_0 = \mu_0 + \tan^{-1}(s_0/c_0) \quad . \qquad (7.104)$$

Note that

$$S_1^2 = -a^2 I$$

and therefore

$$\exp(S_1) \approx I \cos a + S_1 \sin a/a \quad .$$

From this we can easily calculate the elements a_{ij} defined by

$$a_{ij} = [\exp(S_1)]_{ij} \quad .$$

Using these a_{ij}, we construct the S matrix elements from (5.89):

$$S_{11} = \exp(2i n_1)[\cos^2 a - \sin^2 a \, \exp(2i\tau_0)] \qquad (7.105a)$$

$$S_{12} = -i^{\ell_1 - \ell_2 + 1} \exp(i n_1 + i n_2) \sin(2a) \sin\tau_0 \qquad (7.105b)$$

where the phase shifts n_i are defined in terms of the diabatic momentum $p_i(r)$:

$$n_i = \frac{\pi}{2} (\ell + \frac{1}{2}) - k_i r_i + \int_{r_i}^{\infty} dx[\lambda p_i(x) - k_i] \quad .$$

131

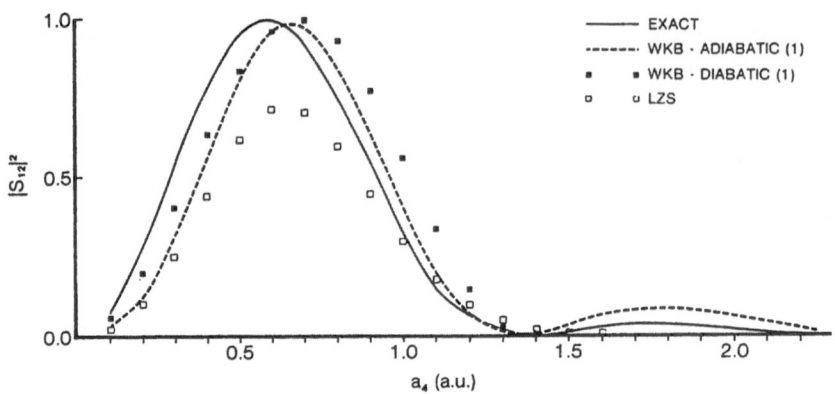

Fig.7.3. Comparison of transition probabilities as a function of coupling constant a_4

Table 7.3. $|S_{12}|^2$ vs coupling strength a_4 for E = 70.9 eV and ℓ = 310. Parameters for potentials are given in Table 7.1.

a_4 [a.u.]	Adiabatic Exact	1	2	Diabatic 1	2	LZS
0.1	0.0732	0.0253	0.0257	0.0517	0.0309	0.0213
0.2	0.2704	0.1223	0.1253	0.1961	0.1209	0.0988
0.3	0.5320	0.3088	0.3142	0.4033	0.2620	0.2450
0.4	0.7813	0.5574	0.5611	0.6305	0.4398	0.4370
0.5	0.9494	0.7976	0.7988	0.8304	0.6328	0.6141
0.6	0.9958	0.9508	0.9534	0.9618	0.8125	0.7115
0.7	0.9177	0.9672	0.9745	0.9974	0.9452	0.6992
0.8	0.7458	0.8465	0.8560	0.9299	0.9978	0.5947
0.9	0.5301	0.6334	0.6382	0.7732	0.9477	0.4442
1.0	0.3219	0.4943	0.3899	0.5597	0.7934	0.2946
1.2	0.0551	0.0592	0.0481	0.1422	0.3111	0.0932
1.4	0.0006	0.0069	0.0155	0.0052	0.0046	0.0192
1.6	0.0249	0.0699	0.0989	0.0000	0.0000	0.0025
2.0	0.0127	0.0540	0.0804			0.0000
2.4	0.0001	0.0049	0.0137			

Calculation 1 includes only the matrix elements of M_∞ which have stationary phase. See (7.68b,105b).
Calculation 2 includes the stationary- as well as nonstationary-phase terms in the matrix M_∞. See (7.58,98).

These forms for the S-matrix elements are similar to those in the adiabatic representation theory discussed in Sect.7.2.2.

The accuracy of the method is assessed by calculating S_{12} given by (7.105) for the curve-crossing model (7.97) in Sect.7.2.5 and comparing the results with the numerical solution transition probabilities (Fig.7.3). Table 7.3 also summarizes the numerical results.

It is also possible to apply the modified Magnus approximation for solving the equation of motion (5.104). Since the structures of the matrices $\mathbf{W}$ and $\mathbf{M}_\infty$ in the equations of motion (5.53,104) are isomorphic to each other, it is possible to obtain the S-matrix elements by simply replacing various quantities in (7.94-96) with the equivalents in the diabatic-representation theory. The correspondences are achieved by the following equivalences:

$$(v - v_t) \to \lambda^2 \int_{r_t}^{r} dx [w_{12}/(p_1 p_2)^{1/2}] \equiv \omega(r)$$

$$\Delta_0 \to \mu_0$$

$$\mathbf{W}_0 \to D[\mu_0]$$

$$(s_1, c_1) \to (s_0, c_0) \quad,$$

and by the definitions:

$$c_i(\infty) = \frac{1}{2} \int_{r_t}^{\infty} dx \, h_i(x) \, \cos\mu(x) \lambda^2 w_{12}(x)/[p_1(x)p_2(x)]^{1/2}$$

$$s_i(\infty) = \frac{1}{2} \int_{r_t}^{\infty} dx \, h_i(x) \, \sin\mu(x) \lambda^2 w_{12}(x)/[p_1(x)p_2(x)]^{1/2} \quad,$$

where $i = 2,3$ and

$$h_2(x) = \cos\omega(x) \quad, \quad h_3(x) = \sin\omega(x) \quad.$$

It is useful to note that

$$\pi/2 - v_t \to \omega(\infty) \quad.$$

Thus we obtain the S-matrix element as follows:

$$S_{12} = -i^{\ell_1 - \ell_2 + 1} 2^{\ell_1 - \ell_2 + 1} \exp(in_1 + in_2)\sin(2\lambda_2)f \quad,$$

where

$$f = \cos[\omega(\infty) - \pi]\cos\xi \, \sin\mu_0 + \sin\xi \, \cos\alpha \, \cos\mu_0$$

$$- \cot(2\lambda_2) \sin[\omega(\infty) - \pi] \sin\mu_0$$

$$- \tan\lambda_2 \, \sin\xi \, \sin\alpha\{\sin\xi \, \cos[\omega(\infty) - \alpha + \pi] \sin\mu_0$$

$$- \cos\xi \, \cos\mu_0\} \quad,$$

$$\xi = \tan^{-1}\{[s_2^2(\infty) + s_3^2(\infty)]^{1/2}/[\omega(\infty)/2 - c_0]\}$$

$$\alpha = \tan^{-1}[s_3(\infty)/s_2(\infty)]$$

$$\lambda_2 = [(\omega(\infty)/2 - c_0)^2 + s_2^2 + s_3^2(\infty)]^{1/2} \quad .$$

Here λ_2 is the eigenvalue of the involution matrix $\mathbf{P}$ defined by

$$P(r) = \frac{1}{2} \int_{r_t}^{r} dx\ \lambda^2 w_{12}[p_1(x)p_2(x)]^{-1/2}\{D[\mu^{(-)}(x)] - D[\mu_0]$$

$$+ \frac{1}{2} \sin\omega(x)[M_\infty, D[\mu_0]]$$

$$+ \sin^2 \frac{1}{2}\omega(x)\ D(\mu_0)[M_\infty^0, M_\infty]\} \quad ,$$

with M_∞^0 defined by

$$M_\infty^0 = \{\lambda^2 w_{12}(r)/2[p_1(r)p_2(r)]^{1/2}\}D[\mu_0] \quad .$$

That is, we have

$$P^2 = -\lambda_2^2 I \quad .$$

One can verify the above formulas by starting from (7.100) for M_∞, replacing $\mathbf{P}$ with P defined above, and following the procedure used for (7.94-96) in Sect. 7.2.4.

8. Curve-Crossing Problems. II: Multistate Models

We continue the discussion of curve-crossing problems in this chapter and generalize the foregoing theories to an n-state problem. It is very difficult to obtain transition probabilities in forms of simplicity comparable to those presented in Chap.7, simply because the equations of motion no longer possess the same useful properties as those for two-state problems unless some further simplifying approximations are introduced. It is too much to hope at this time for a general analytic theory. This certainly is an area in semiclassical scattering theory where further work is required. In this chapter we shall consider a curve-crossing model [8.1-6] which yields quite simple analytic results. Perhaps and hopefully, one may in the future be able to build a better and more general theory with the clues and insights provided by the study of a model of this kind.

8.1 Multichannel Curve-Crossing Problems

To be specific, we shall consider a case where n-1 repulsive potential curves cross an attractive potential (Fig.8.1). The repulsive potentials (electronic states, for example) are numbered 1 through n-1 and the attractive potential is numbered n. The model is representative of various multi-curve-crossing models considered by various authors [8.1-6]. In most of the papers available [8.1-5] the Landau-Zener-Stückelberg theory or its variations are utilized. The basic idea in all these theories is that a multi-state curve-crossing problem can be decomposed into a string of local two-state curve-crossing problems if the curve-crossing points are sufficiently separated. Unfortunately, a clear-cut criterion does not appear to be available on how far apart the curve-crossing points must be before this kind of theory is applicable, but a rough rule of thumb would be that their separations should be much larger than the wavelength of the relative motion. Since we have a well-defined mathematical procedure already developed in the form of

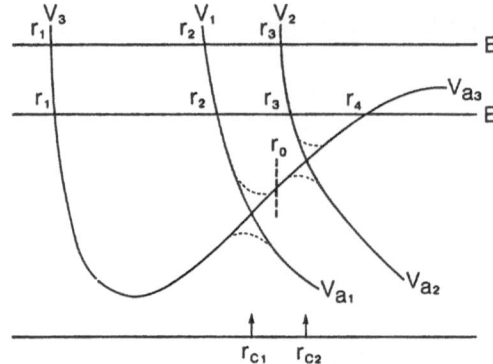

Fig.8.1. A three-curve-crossing problem. The potential curves in solid line are diabatic potentials and the dotted lines joining the solid lines represent the adiabatic potentials

the uniform semiclassical theory which gives as simple results as the LZS theory, but performs equally well or better than the latter, we shall base our discussion on the uniform semiclassical adiabatic representation approach [5.14, 8.6-9].

The equation of motion is given by

$$\frac{d}{dr} X(r) = W(r)X(r) \quad , \tag{8.1}$$

where the matrix W is defined in (5.78). Working out the matrix elements more explicitly, we find from (5.73,75,78) that W has the form

$$
W(r) = \begin{pmatrix}
0 & \hat{F}_{12}P_{21} & \cdots & \hat{F}_{1n}P_{n1} \\
\hat{F}_{21}P_{12} & 0 & \cdots & \hat{F}_{2n}P_{n2} \\
\hat{F}_{31}P_{13} & \hat{F}_{32}P_{23} & \cdots & \hat{F}_{3n}P_{n3} \\
\vdots & \vdots & & \vdots \\
\hat{F}_{n1}P_{1n} & \hat{F}_{n2}P_{2n} & \cdots & 0
\end{pmatrix} \tag{8.2}
$$

where

$$\hat{F}_{ij} = \sum_{k=1}^{n} H_{ki}h'_{kj} \quad .$$

We shall specialize first to a three-state problem to illustrate the method and then generalize the result to the n-state situation. Therefore, there are two curve-crossing points to consider. If they are sufficiently well separated from each other, then in the vicinity of the first curve-crossing point r_{c1} where a transition between states 1 and 2 occurs the

136

third state has little influence. Consequently, $\hat{F}_{ij}$ may be approximated as follows:

$$\hat{F}_{12} = -\hat{F}_{21} \approx \frac{1}{2} (1 + u_1^2)^{-1} \frac{du_1}{dr}$$

$$\hat{F}_{23} \approx 0 \qquad\qquad (8.3)$$

$$\hat{F}_{31} \approx 0 \quad .$$

Similarly, in the vicinity or r_{c2} we may put

$$\hat{F}_{12} \approx 0$$

$$\hat{F}_{23} = -\hat{F}_{32} \approx \frac{1}{2} (1 + u_2^2)^{-1} \frac{du_2}{dr} \qquad\qquad (8.4)$$

$$\hat{F}_{31} \approx 0 \quad .$$

In (8.3,4)

$$u_1 = [E_1(r) - E_3(r)]/2w_{13} \qquad\qquad (8.5a)$$

$$u_2 = [E_2(r) - E_3(r)]/2w_{23} \quad . \qquad\qquad (8.5b)$$

Note that the states in (8.5a,b) are numbered respectively (1,3) and (2,3) which label the diabatic potentials. This is due to the fact that the matrix **W** is ordered according to the adiabatic states that do not coincide everywhere with the diabatic states.

The physical reason behind (8.3,4) is as follows. The three-state model requires solution of (5.1) for n = 3, i.e., 3-coupled radial equations. If the coupling potentials are localized at the crossing points, these coupled equations can be approximated at each crossing point by a pair of two-coupled equations and an equation that is decoupled from the pair. That is, the three-state problem is effectively reduced to a two-state and one-state problem. Then it is now possible to apply a theory developed for two-state problems. Equations (8.3,4) are the results of applying the two-state uniform semiclassical theory (Sect.5.3) to the local two-state problems in the model assumed here.

Equation (8.3) implies that the equation of motion is given in the form

$$\frac{d}{du_1} X(u_1) = \frac{1}{2} (1 + u_1^2)^{-1} \mathbf{W}_1(u_1) X(u_1) \qquad\qquad (8.6)$$

$$W_1(u_1) = \begin{pmatrix} 0 & P_{21} & 0 \\ -P_{12} & 0 & 0 \\ 0 & 0 & 0 \end{pmatrix} \qquad\qquad (8.7)$$

which is a 6×6 matrix, while (8.4) implies the equation of motion in the form

$$\frac{d}{du_2} X(u_2) = \frac{1}{2} (1 + u_2^2)^{-1} W_2(u_2) X(u_2) \qquad (8.8)$$

$$W_2(u_2) = \begin{pmatrix} 0 & 0 & 0 \\ 0 & 0 & P_{32} \\ 0 & -P_{23} & 0 \end{pmatrix} . \qquad (8.9)$$

Since the equation of motion is a first-order differential equation, the solution X in the present sequential pairwise crossing approximation consists of two factors determined by W_1 and W_2 respectively. The implication of this approximation is clearly seen with the product integral form (5.93) for the solution of (8.1), which splits into two factors

$$X(\infty) = \prod_{r_0}^{\infty} \exp[W(s)ds] X_0$$

$$= \prod_{r_m}^{\infty} \exp[W(s)ds] \prod_{r_0}^{r_m} \exp[W(s)ds] X_0$$

$$= \prod_{r_m}^{\infty} \exp[W_2(s)ds] \prod_{r_0}^{r_m} \exp[W_1(s)ds] X_0 \qquad (r_0 < r_m < \infty) , \qquad (8.10)$$

where $X_0 = X(r_0)$ is the initial condition

$$X_0 = \{\alpha_1^0, 0, \alpha_2^0, 0, \alpha_3^0, 0\} .$$

Especially in the first-order Magnus approximation the product integral (8.10) may be written

$$X(\infty) = \exp(f_2) \exp(f_1) X_0 \qquad (8.11)$$

where the matrices f_1 and f_2 are defined by

$$f_i = \frac{1}{2} \int_{u_{i1}}^{u_{i2}} du (1 + u^2)^{-1} W_i(u) , \qquad (i = 1,2) \qquad (8.12a)$$

with u_{i1} and u_{i2} denoting appropriate end points that may be defined by the turning points or a suitable midpoint between r_{c1} and r_{c2} or infinity. The matrix f_i has the same properties as F_1 in Chap.7. In particular,

138

$$f_1^2 = -\lambda_1^2 \begin{pmatrix} I & 0 \\ 0 & 0 \end{pmatrix} , \qquad (8.12b)$$

where I is the four-dimensional unit matrix and

$$\lambda_1^2 = I(A_1,B_2) \ I(A_2,B_1) - I(A_1,A_2) \ I(B_2,B_1) .$$

Here the notations are the same as in Chap.7. Similar definitions apply for λ_2 corresponding to f_2. By using (8.12b), it is easy to show that

$$\exp(f_1) = \begin{pmatrix} \bar{c}_1 I & \bar{s}_1 f_{21} & 0 \\ -\bar{s}_1 f_{12} & \bar{c}_1 I & 0 \\ 0 & 0 & I \end{pmatrix} \qquad (8.13)$$

where

$$f_{ij} = \frac{1}{2} \int_{u_{i1}}^{u_{i2}} du(1 + u^2)^{-1} P_{ij}(u) \qquad (8.14)$$

$$\bar{c}_i = \cos\lambda_i$$

$$\bar{s}_i = \sin\lambda_i/\lambda_i ,$$

and similarly for $\exp(f_2)$. When these results are put together, we obtain from (8.11) the solution $X(\infty)$ at r infinite in the following form:

$$X(\infty) = \begin{pmatrix} I & 0 & 0 \\ 0 & \bar{c}_2 I & \bar{s}_2 f_{32} \\ 0 & -\bar{s}_2 f_{23} & \bar{c}_2 I \end{pmatrix} \begin{pmatrix} \bar{c}_1 I & \bar{s}_1 f_{21} & 0 \\ -\bar{s}_1 f_{12} & \bar{c}_1 I & 0 \\ 0 & 0 & I \end{pmatrix} X_0 \qquad (8.15)$$

$$\equiv a X_0 .$$

Now the transition probability amplitudes can be calculated from the matrix **a** in (8.15). The integrals appearing in the matrices f_{ij} can be computed numerically and their computation can proceed in the same manner as for the two-state problems already discussed — in an approximation by applying the stationary-phase method. Since the details are available in the literature — and it is not difficult to verify them — we shall simply present the essential results below. First define

$$\tau_i = \lambda \int_{r_i}^{r_{ci}} \Lambda_i \ dx - \lambda \int_{r_{i+1}}^{r_{ci}} \Lambda_{i+1} \ dx \qquad (8.16)$$

139

$$\psi_i^{(-)}(r) = \lambda \int_{r_{ci}}^{r} (\Lambda_i - \Lambda_{i+1})dx \qquad (8.17)$$

$$c_i = \frac{1}{2} \int_{u_{i1}}^{u_{i2}} du(1 + u^2)^{-1} \cos\psi_i^{(-)}(u)$$

$$s_i = \frac{1}{2} \int_{u_{i1}}^{u_{i2}} du(1 + u^2)^{-1} \sin\psi_i^{(-)}(u) \qquad (8.18)$$

$$P_{ij}(\tau_i) \approx R_{ij}(\tau_i) = \begin{pmatrix} \cos\tau_i & \sin\tau_i \\ -\sin\tau_i & \cos\tau_i \end{pmatrix} , \qquad (i < j) . \qquad (8.19)$$

Then the matrices W_i take the form

$$W_i = \frac{1}{2} \left[D_i(\tau_i)\cos\psi_i^{(-)} + D_i(\tau_i + \pi/2)\sin\psi_i^{(-)} \right](1 + u_i^2)^{-1}(du_i/dr) \qquad (8.20a)$$

where

$$D_1(\tau_1) = \begin{pmatrix} 0 & R_{21}(\tau_1) & 0 \\ -R_{12}(\tau_1) & 0 & 0 \\ 0 & 0 & 0 \end{pmatrix} \qquad (8.20b)$$

$$D_2(\tau_2) = \begin{pmatrix} 0 & 0 & 0 \\ 0 & 0 & R_{32}(\tau_2) \\ 0 & -R_{23}(\tau_2) & 0 \end{pmatrix} \qquad (8.20c)$$

and thus the matrix f_i is given in terms of the matrix $D_i(\tau_i)$:

$$f_i = D_i[\tau_i]c_i + D_i[\tau_i + \pi/2]s_i$$

$$= \lambda_i D_i[\tau_i + \gamma_i]$$

$$\equiv \lambda_i D_i[\varepsilon_i] \qquad (8.21a)$$

where

$$\lambda_i = (c_i^2 + s_i^2)^{1/2} ,$$

$$\gamma_i = \tan^{-1}(s_i/c_i) ,$$

$$\varepsilon_i = \tau_i + \gamma_i . \qquad (8.21b)$$

These notations are completely parallel to those in Chap.7. Notice also that the form of f_i is very similar to F_1 in the stationary-phase method; see (7.66). These relations show that

$$f_{12} = R_{12}(\tau_1) \quad , \quad f_{21} = R_{21}(\tau_1)$$

$$f_{23} = R_{23}(\tau_2) \quad , \quad f_{32} = R_{32}(\tau_2) \quad .$$

By using the above results in (8.15) and (5.89), we construct the S-matrix elements for the three-state problem:

$$S_{21} = -\exp[i(\eta_1 + \eta_2)][(1 + \omega_1)(1 + \omega_2)]^{-1}[\omega_1(1 + \omega_2)]^{1/2} \sin\tau_1 \qquad (8.22a)$$

$$S_{22} = \exp(2i\eta_2)[(1 + \omega_1)(1 + \omega_2)]^{-1}[1 + \omega_1 \exp(2i\tau_1) + \omega_2 \exp(-2i\tau_2)$$

$$+ \omega_1\omega_2 \exp(-2i\tau_2)] \qquad (8.22b)$$

$$S_{23} = -\exp[i(\eta_2 + \eta_3)][(1 + \omega_1)(1 + \omega_2)]^{-1}\omega_2^{1/2}$$

$$\times [\omega_1 \exp(i\tau_1) \sin(\tau_1 + \tau_2) + \sin\tau_2] \quad , \qquad (8.22c)$$

where

$$\omega_i = 1/\lambda_i \quad . \qquad (8.22d)$$

The transition-probability amplitudes above are reminiscent of the results by *Dubrovskii* [8.10] who used the LZS theory.

8.2 The Case of n > 3

The three-state results above can be generalized to an n-state problem where (n-1) repulsive curves cross pairwise with the nth curve (Fig.8.2). Formally, the generalization is almost trivial, but the n-state problem for n > 3 requires the solution of a certain matrix algebraic problem, if explicit formulas similar to (8.22) are sought for the S-matrix elements.

Since the potential curves cross pairwise, the matrix $\textbf{\textit{W}}$ can be broken up into a sequence of two-state forms. More explicitly put, with the definition of four-dimensional matrices,

$$G_i = \begin{pmatrix} \bar{c}_i I & \bar{c}_i f_{i+1,i} \\ -\bar{s}_i f_{i,i+1} & \bar{c}_i I \end{pmatrix} \quad , \qquad (8.23)$$

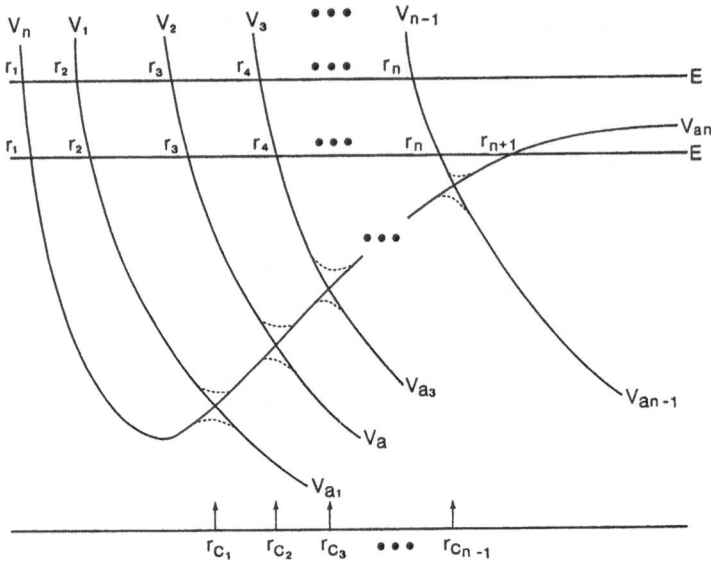

Fig.8.2. An N curve-crossing problem. The meanings of the symbols and lines are given in Fig.8.1

which may be approximated in the stationary-phase method by

$$G_i = \begin{pmatrix} \cos\lambda_i I & \sin\lambda_i R_{i+1,i}(\tau_i) \\ -\sin\lambda_i R_{i,i+1}(\tau_i) & \cos\lambda_i I \end{pmatrix} \quad , \qquad (8.24)$$

we obtain an approximate solution $X(\infty)$ in the form

$$X(\infty) = \begin{pmatrix} I & & & 0 \\ & I & & \\ & & \ddots & \\ 0 & & & G_{n-1} \end{pmatrix} \begin{pmatrix} I & & & 0 \\ & I & & \\ & & \ddots & \\ 0 & & & G_{n-2} \\ & & & & I \end{pmatrix} \cdots \begin{pmatrix} G_1 & & & 0 \\ & I & & \\ & & \ddots & \\ 0 & & & I \end{pmatrix} X_0$$

$$\equiv a X_0 \quad . \qquad (8.25)$$

The S-matrix elements then can be constructed from the matrix a above. If the initial state is designated the $(n-1)$th state, then the S-matrix elements are given in the form

$$S_{n-1,k} = \exp[i(\eta_{n-1} + \eta_k)] \sum_{j=1}^{n} (a_{2k-1,j} + i a_{2k,j}) \Delta_{n-1,j}/\Delta \quad , \qquad (8.26a)$$

where Δ is a determinant whose elements are given in terms of the elements of the matrix a in (8.25):

142

$$a_{2i-1,2j-1} - ia_{2i,2j-1}$$

(8.26b)

and $\Delta_{n-1,j}$ is the $(n-1,j)$ cofactor of Δ.

Therefore, it is now necessary to find the elements a_{ij} explicitly from the matrix **a** in order to construct the determinant and then obtain its cofactors. This is simple and straightforward if n is small, but not trivial for large n. It is a linear algebraic problem. The solution is given in the form of rules without proof.

Let $c_0 = c_n = 1$.

Rule 1. The jth element d_{ji} of the ith column of Δ is given by

$$d_{ji} = c_{i-1}c_j \prod_{k=1}^{j-1} [-s_k \exp(-i\tau_k)] \quad , \quad (i = 1,2,\ldots,n) \quad ; \tag{8.27}$$

the product must be put equal to 1 when the upper limit is less than the lower limit.

Rule 2. For $j < i$

$$d_{i-1,i} = s_{i-1} \exp(i\tau_{i-1}) \quad , \quad (i = 1,2,\ldots,n) \quad ,$$

$$d_{ji} = 0 \quad \text{for all} \quad j < i-1 \quad . \tag{8.28}$$

The determinant is then

$$\Delta = \det|d_{ji}| \quad . \tag{8.29}$$

To find the transition amplitudes more explicitly, it is necessary to construct an auxiliary determinant.

Rule 3. Take the complex conjugate of d_{ji}. Then the auxiliary determinant $\Delta^{(*)}$ is given by

$$\Delta^{(*)} = \det|d_{ij}^*| = \Delta^* \quad . \tag{8.30}$$

The transition amplitudes are now given as follows:

$$T_{n-i,j} + (2i)^{-1}\delta_{j,n-1} = (2i)^{-1} \exp[i(n_{n-1} + n_j)]$$

$$\times \sum_{k=1}^{n} \Delta_{n-1,k} d_{jk}^*/\Delta \quad (j = 1,2,\ldots,n) \quad . \tag{8.31}$$

This is the result we set out to find for the inelastic scattering of the n curve-crossing problem. The rules above can be verified by explicit calculations for lower n's and may be shown to be true by induction. As an example, we show the determinant for $n = 4$ constructed with the rules:

$$\Delta = \begin{pmatrix} c_1 & s_1 \exp(i\tau_1) & 0 & 0 \\ -c_2 s_1 \exp(-i\tau_1) & c_1 c_2 & s_2 \exp(i\tau_2) & 0 \\ c_3 s_1 s_2 \exp(-i\tau_{12}) & -c_1 c_3 s_2 \exp(-i\tau_2) & c_2 c_3 & s_3 \exp(i\tau_3) \\ -s_1 s_2 s_3 \exp(-i\tau_{123}) & c_1 s_2 s_3 \exp(-i\tau_{23}) & -c_2 s_3 \exp(-i\tau_3) & c_3 \end{pmatrix}$$

where

$$\tau_{12\ldots\ell} = \tau_1 + \tau_2 + \ldots + \tau_\ell \quad , \tag{8.32}$$

with τ_i defined by (8.16).

9. Curve-Crossing Problems. III: Predissociations

In predissociation phenomena [9.1] curve-crossings also play an important role. Experimental measurements on level shifts and widths due to predissociation can provide valuable information on the interaction potential parameters for systems undergoing a predissociation [9.2-4]. Predissociation was studied semiclassically by different methods. In the method developed by *Child* and *Bandrauk* [9.5,6] and their collaborators [8.5,9.7,8], the potential functions are approximated by linear functions around the curve-crossing point and then the associated two-state problem is solved exactly in the momentum representation. The method yields the transition probabilities similar to the LZS theory as expected, but requires somewhat arbitrary fittings of the potential curves after the transition probabilities are obtained for the locally linear potential problem. In another method, *Dubrovskii* and *Fischer-Hjalmars* [9.9] use a two-state impact parameter method in which two coupled linear first-order equations are converted into a pair of second-order equations which are then solved similarly to Zener's classic method already mentioned. Their method also gives the transition probabilities comparable to those from the LSZ theory except for an additional phase factor similar to that by *Child* [9.5]. These two are basically in the same line of approach as the LZS theory, since the linear approximation is made to the potentials near the curve crossing point. If the phase correction factor were ignored, one could recover their results with the Stückelberg method by proceeding in the same manner as in Sect.7.1. In still another method one can use a uniform semiclassical method [9.10,8.6] in much the same spirit as in the theory developed for inelastic scattering in Chap.5. Since the uniform semiclassical method is generally more accurate than the LZS theory for inelastic scattering, it is expected to be the case with predissociation. The present chapter is devoted to a uniform semiclassical theory of predissociation. For the space limitation we shall only consider the case of energy higher than the minimum of the upper adiabatic potential or, roughly speaking, the curve crossing energy. Other cases can be studied similarly and the reader is referred to the literature [8.6,9.10].

9.1 A Two-State Model

Let us consider a case of a repulsive potential crossing an attractive potential that can accommodate bound states (Fig.8.1). In the literature a distinction is made between the cases where the crossing occurs at the outer attractive limb and at the inner repulsive limb of the potential. They are called the outer and inner crossing case, respectively. They must be distinguished, basically because either the potentials are approximated by linear functions or an equivalent approximation is introduced during the analysis in an attempt to obtain some analytic results. In the present approach there is no necessity to make such a distinction.

The equations to be solved are formally the same as (5.1) for n = 2. Therefore, at a quick glance at the problem one might be tempted to think that the theory will be the same as in Sect.5.3. However, there is an important difference because it also involves a two-turning-point problem, which must be treated properly. In fact, it is the key to a theory that unifies the outer and inner crossing models appearing in other theories mentioned. We have already dealt with a similar problem in Sect.4.3, where a three-turning-problem was solved in terms of two-turning-point problems. The differential equations for wave functions involved here are the same as those in Sect.4.3.

Let us call the adiabatic potentials V_{a1} and V_{a2}, the latter being the upper adiabatic potential. Here we shall assume that the energy is higher than the minimum V_{min} of V_{a2}.

The wave functions for the two-turning-point problem associated with V_{a2} is determined by the differential equation

$$\left[d^2/dr^2 + \lambda^2 f_2(\phi_2)\phi_2'^2 + \frac{1}{2} \{\phi_2,r\} \right] Y_2 = 0 \tag{9.1}$$

with

$$f_2(\phi_2) = C^2 - \phi_2^2 \quad . \tag{9.2}$$

The meanings of C and $\phi_2(r)$ are the same as in (4.58-60). The constant C will turn out to be the Bohr-Sommerfeld quantization condition for the bound states of V_{a2},

$$\frac{1}{2} \lambda C \pi = (n + \frac{1}{2})\pi = \lambda \int_{r_1}^{r_2} dr \, \Lambda_2(r) \quad , \tag{9.3}$$

and $\phi_2(r)$ is the new variable in terms of which the uniform semiclassical wave function is expressed; see (4.59). With transformation

$$Y_2 = (\phi_2')^{-1/2}W_2 \qquad (9.4)$$

we again find that W_2 satisfies the differential equation for parabolic cyclinder functions. We shall denote the two linearly independent solutions by $U(-n-\frac{1}{2};-\xi)$ and $V(-n-\frac{1}{2};-\xi)$, using the standard notations, and therefore the two independent solutions for (9.1) by

$$A_2 = N(2\pi)^{-1/4}(\xi')^{-1/2}U(-n-\frac{1}{2};-\xi)$$

$$= N(2\pi)^{-1/4}(\xi')^{-1/2}D_n(-\xi) \qquad (9.5)$$

$$B_2 = N^{-1}(2\pi)^{-1/4}(\xi')^{-1/2}V(-n-\frac{1}{2};-\xi)$$

$$= N^{-1}(2\pi)^{-1/2}\pi^{-1}\Gamma(-n)[-\cos n\pi \, D_n(-\xi) + D_n(\xi)] \quad , \qquad (9.6)$$

where

$$N = \Gamma(n + 1)^{-1/2}$$

$$\xi = (2\lambda)^{1/2}\phi_2$$

and the prime on ξ and ϕ_2 denotes differentiation with respect to r.

The uniform semiclassical wave functions for the adiabatic potential V_{a1} are the same as those in the case of a single-turning-point problem in the previous chapters. They are given in terms of the Airy functions:

$$A_1 = (t')^{-1/2}\text{Ai}(-t) \qquad (9.7)$$

$$B_1 = (t')^{-1/2}\text{Bi}(-t) \quad , \qquad (9.8)$$

where

$$t = \lambda^{2/3}\phi_1$$

with ϕ_1 defined by (5.46b). The prime means differentiation with respect to r.

With this preparation, we can write the wave function u_i in the form

$$u_i = \sum_{j=1}^{2} (\alpha_{ij}A_j + \beta_{ij}B_j) \qquad (9.9)$$

and proceed in exactly the same manner as for (5.53) for inelastic scattering. In fact, the structure of the resulting equations for α_{11}, α_{22}, β_{11} and β_{22} is exactly the same as (5.53) except for the fact that A_2 and B_2 appearing in the matrix W must be understood by (9.5,6). Therefore, we have the

147

equation of motion

$$\frac{d}{du} X(u) = \frac{1}{2} (1 + u^2)^{-1} w(u) X(u) \quad , \tag{9.10}$$

where

$$X(u) = \{1 + [u - (1 + u^2)^{1/2}]^2\}^{1/2} \{\alpha_{11}, \beta_{11}, \alpha_{22}, \beta_{22}\} \quad .$$

This equation must be solved subject to the initial condition at $r = 0$ or $u = -\infty$:

$$X_0 = X(u = -\infty) = \{\alpha_1^0, 0, \alpha_2^0, 0\}$$

and also the boundary conditions at $r = \infty$ such that the resulting wave functions are finite or satisfy the scattering boundary condition. We express the solution of (9.10) in the form

$$X(\infty) = aX_0 \tag{9.11}$$

as before. That is, with the initial condition just stated, the solutions may be written as

$$\alpha_{11}(\infty) = a_{11}\alpha_1^0 + a_{13}\alpha_2^0$$

$$\beta_{11}(\infty) = a_{21}\alpha_1^0 + a_{23}\alpha_2^0 \tag{9.12}$$

$$\alpha_{22}(\infty) = a_{31}\alpha_1^0 + a_{33}\alpha_2^0$$

$$\beta_{22}(\infty) = a_{41}\alpha_1^0 + a_{43}\alpha_2^0 \quad .$$

Here a_{ij} are the matrix elements of a in (9.11) which can be found by solving the equation of motion (9.10). The boundary conditions on u_1 and u_2 at large r are

$$u_1 \sim i^\ell \sin(kr - \frac{1}{2} \ell\pi) + T_{11} \exp(ikr) \quad ,$$

$$u_2 \sim 2^{1/2} T_{12} \exp(-|k|r) \quad , \tag{9.13}$$

where we have assumed the orbital angular momenta are the same for both states.

Expressing the asymptotic formulas for A_j and B_j as follows

$$A_1 \sim (\pi k)^{-1/2} \left[c_{a1}^{(+)} \exp(ikr) + c_{a1}^{(-)} \exp(-ikr) \right]$$

$$B_1 \sim (\pi k)^{-1/2} \left[c_{b1}^{(+)} \exp(ikr) + c_{b1}^{(-)} \exp(-ikr) \right]$$

$$A_2 \sim (\pi k)^{-1/2}\left[c_{a2}^{(-)} \exp(-|k|r) + c_{a2}^{(+)} \exp(|k|r)\right]$$

$$B_2 \sim (\pi k)^{-1/2}\left[c_{b2}^{(-)} \exp(-|k|r) + c_{b2}^{(+)} \exp(|k|r)\right] , \qquad (9.14)$$

substituting into (9.9) together with (9.12), and comparing the asymptotic wave functions with (9.13), we find the transition amplitudes

$$T_{11} + (2i)^{-1} = (\pi k)^{-1/2}\left[\left(c_{a1}^{(+)}a_{11} + c_{b1}^{(+)}a_{21}\right)\alpha_1^0 + \left(c_{a1}^{(+)}a_{13} + c_{b1}^{(+)}a_{23}\right)\alpha_2^0\right] \qquad (9.15)$$

$$T_{12} = (2/\pi|k|)^{1/2}\left[\left(c_{a2}^{(-)}a_{31} + c_{b2}^{(-)}a_{41}\right)\alpha_1^0 + \left(c_{a2}^{(-)}a_{33} + c_{b2}^{(-)}a_{43}\right)\alpha_2^0\right] , \qquad (9.16)$$

where the constants α_1^0 and α_2^0 are determined from the linear equations

$$\frac{1}{2} i^{2\ell+1}(\pi k)^{1/2} = \left(c_{a1}^{(-)}a_{11} + c_{b1}^{(-)}a_{21}\right)\alpha_1^0 + \left(c_{a1}^{(-)}a_{13} + c_{b1}^{(-)}a_{23}\right)\alpha_2^0 \qquad (9.17)$$

$$0 = \left(c_{a2}^{(+)}a_{31} + c_{b2}^{(+)}a_{41}\right)\alpha_1^0 + \left(c_{a2}^{(+)}a_{33} + c_{b2}^{(+)}a_{43}\right)\alpha_2^0 . \qquad (9.18)$$

The solution of these equations yields the constants

$$\alpha_1^0 = \frac{1}{2} i^{2\ell+1}(\pi k)^{1/2}\Delta^{-1}\left(c_{a2}^{(+)}a_{33} + c_{b2}^{(+)}a_{43}\right)$$

$$\alpha_2^0 = -\frac{1}{2} i^{2\ell+1}(\pi k)^{1/2}\Delta^{-1}\left(c_{a2}^{(+)}a_{31} + c_{b2}^{(+)}a_{41}\right) , \qquad (9.19)$$

where

$$\Delta = \left(c_{a1}^{(-)}a_{11} + c_{b1}^{(-)}a_{21}\right)\left(c_{a2}^{(+)}a_{33} + c_{b2}^{(+)}a_{43}\right)$$

$$- \left(c_{a1}^{(-)}a_{13} + c_{b1}^{(-)}a_{23}\right)\left(c_{a2}^{(!)}a_{31} + c_{b2}^{(+)}a_{41}\right) . \qquad (9.20)$$

Now by substituting these constants into (9.15,16), we finally obtain T_{11} and T_{12} in terms of dynamical quantities alone, i.e., a_{ij}, $c_{ai}^{(\pm)}$, and $c_{bi}^{(\pm)}$.

The coefficients $c_{ai}^{(\pm)}$ and $c_{bi}^{(\pm)}$ can easily be found from the asymptotic formulas for the Airy functions and the parabolic cylinder functions (App.1):

$$c_{a1}^{(\pm)} = \frac{1}{2} i^{\mp(\ell+1)} \exp(\pm i n_1)$$

$$c_{b1}^{(\pm)} = \frac{1}{2} i^{\mp\ell} \exp(\pm i n_1) \qquad (9.21)$$

and with $a \equiv n + \frac{1}{2}$

149

$$c_{a2}^{(+)} = (2/\pi)^{1/4} N\Gamma(a + \tfrac{1}{2})(e/a)^{a/2} \exp(n_2) \cos a\pi$$

$$c_{a2}^{(-)} = -i(\pi/2)^{1/4} N(a/e)^{a/2} \exp(-n_2 + i\pi a)$$

$$c_{b2}^{(+)} = -(2/\pi)^{1/4} N^{-1}(e/a)^{a/2} \exp(n_2) \sin a\pi \qquad\qquad (9.22)$$

$$c_{b2}^{(-)} = \pi^{-1/2}(2\pi)^{-1/4} N^{-1}\Gamma(\tfrac{1}{2} - a)(a/e)^{a/2} \exp(-n_2 + ia\pi) \cos a\pi \quad ,$$

where

$$n_2 = - |k|r_3 + \int_{r_3}^{\infty} (\lambda|\Lambda_2(x)| - |k|)dx \quad . \qquad\qquad (9.23)$$

By taking the Stirling formula for the gamma functions, which is consistent with the semiclassical approximation taken in the present theory, we may simplify (9.22) considerably:

$$c_{a2}^{(+)} = \sqrt{2} \exp(n_2) \cos a\pi$$

$$c_{a2}^{(-)} = -i\, 2^{-1/2} \exp(-n_2) \exp(ia\pi)$$

$$c_{b2}^{(+)} = -\sqrt{2} \exp(n_2) \sin a\pi$$

$$c_{b2}^{(-)} = 2^{-1/2} \exp(-n_2) \exp(ia\pi) \quad . \qquad\qquad (9.24)$$

The matrix **a** must now be determined. Here we use the first-order Magnus approximation for X,

$$X(u = u_3) = \exp(\mathbf{F})X_0$$
$$= [\cosh f\, \mathbf{I} + (\sinh f/f)\mathbf{F}]X_0 \quad , \qquad\qquad (9.25)$$

where

$$\mathbf{F} = \tfrac{1}{2} \int_{u_2}^{u_3} du(1 + u^2)^{-1} \boldsymbol{w}(u) \qquad\qquad (9.26)$$

$$f = F_{13}F_{31} + F_{14}F_{41} \quad ,$$

with F_{ij} denoting the elements of $\mathbf{F}$, and u_2 and u_3 defined by

$$u_2 = u(r = r_2) \quad , \qquad u_3 = u(r = r_3) \quad .$$

It is useful to recall here the basic properties of the matrix $\boldsymbol{w}(u)$ which are used for the second line in (9.25) [cf. (5.48,51) and p.64].

Note that the upper limit of the integral in (9.26) is u_3, not infinity. It reflects the fact that since one of the wave functions A_2 and B_2 is irregular and increases exponentially for $r > r_3$, it is necessary to cut off the integration range at r_3 to suppress the irregular solution in the domain cut off and thereby obtain the solution satisfying the boundary condition at r infinite.

With the approximate solution and the definitions

$$\theta = -\tan^{-1}(F_{23}/F_{13}) \tag{9.27}$$

$$\psi = \tan^{-1}(F_{41}/F_{31}) \tag{9.28}$$

$$\chi = f^{-2} \tanh^2 f[(F_{31}^2 + F_{41}^2)(F_{13}^2 + F_{23}^2)]^{1/2} \quad , \tag{9.29}$$

$$\beta = a\pi \quad ,$$

we finally find the transition probability amplitudes as follows:

$$T_{11} + (2i)^{-1} = (2i)^{-1} \exp(2i\eta_1) \frac{\cos\beta + \chi \exp(-i\theta)\cos(\beta + \psi)}{\cos\beta + \chi \exp(i\theta) \cos(\beta + \psi)} \tag{9.30}$$

$$T_{12} = -i^\ell \exp(-\eta_2 + i\eta_1) \frac{f^{-1} \tanh f (F_{31}^2 + F_{41}^2)^{1/2} \sin\psi}{\cos\beta + \chi \exp(i\theta)\cos(\beta + \psi)} \quad . \tag{9.31}$$

These formulas are quite similar in structure to those by other authors [9.5,6]. They may be put in more familiar forms if further approximations are made to the quantities appearing in the formulas.

If the stationary-phase method used previously is applied to evaluate the integrals F_{ij}, it can be shown that

$$F_{23} = F_{41} \simeq - J \sin\tau$$

$$F_{31} \simeq - J \cos\tau$$

$$F_{13} \simeq J \cos\tau$$

$$\chi = \tan^2 J \tag{9.32a}$$

where

$$J = \frac{1}{2} \int_{u_2}^{u_3} du(1 + u^2)^{-1} \cos\psi^{(-)} \tag{9.32b}$$

$$\psi^{(-)} = \lambda \int_{r_c}^{r} dx [\Lambda_1(x) - \Lambda_2(x)] \tag{9.32c}$$

$$\tau = \lambda \int_{r_1}^{r_c} dx \, \Lambda_1(x) - \lambda \int_{r_2}^{r_c} dx \, \Lambda_2(x) \quad . \tag{9.32d}$$

Note that these quantities have appeared in the two-state theory considered in Chap.7. To obtain the approximations above for the integrals F_{ij}, we have used the asymptotic formulas of the Airy and parabolic cylinder functions and also neglected the nonstationary-phase parts of the integrals involved. In the same approximation we find the approximations for the phases θ and ψ:

$$\theta = \tau \; ; \quad \psi = \tau \quad .$$

When these results are inserted into (9.30,31), the transition probability amplitudes follow in more familiar forms,

$$T_{11} + (2i)^{-1} = (2i)^{-1} \exp(2in_1) \frac{\cos\beta + \chi^2 \exp(-i\tau)\cos(\beta + \tau)}{\cos\beta + \chi^2 \exp(i\tau) \cos(\beta + \tau)}$$

$$= (2i)^{-1} \exp(2in_1) \frac{\cos(\beta + \tau) + \chi^{-2} \exp(i\tau)\cos\beta}{\cos(\beta + \tau) + \chi^{-2} \exp(i\tau)\cos\beta} \tag{9.33a}$$

$$T_{12} = i^{\ell} \exp(-n_2 + in_1) \frac{\chi \sin\tau}{\cos\beta + \chi^2 \exp(i\tau)\cos(\beta + \tau)} \tag{9.33b}$$

$$= i^{\ell} \exp(-n_2 + in_1) \frac{\chi^{-1} \exp(-i\tau)\sin\tau}{\cos(\beta + \tau) + \chi^{-2} \exp(-i\tau)\cos\beta} \quad .$$

These become exactly the same formulas for the transition probabilities as for the inner crossing model in the theory of *Bandrauk* and *Child* [9.6], if χ^{-2} is replaced with the factor

[exp(2s) - 1]

appearing in the LZS theory; see (7.38) and related expressions in Sect.7.1 . *Child* [9.5] claimed that the S-matrix elements for the outer crossing model are not the same as for the inner crossing model. The present theory shows that the basic nature of the wave functions remains unchanged whether the curve crossing occurs at the inner or outer limb of the potential and therefore there is no compelling physical reason to make such a distinction. Another independent method that uses 'Stückelberg tracing' does not support the assertion either. A careful inspection of the Child-Bandrauk theory shows that such a distinction is necessary due to the use of the diabatic representation and linear approximation to the potentials which makes the approximation for the wave function valid locally only in the neighborhood of the curve-crossing point. But a globalization of the wave functions is required in the end, necessitating a dinstinction between the two cases. The present theory requires none of that. The transition amplitudes obtained here are valid for both inner and outer crossing models of predissociation.

9.2 Modified Magnus Approximation

The transition probability amplitudes in Sect.9.1 are obtained in terms of the first-order Magnus approximation to X. This approximation can be improved by using the modified Magnus approximation discussed in connection with inelastic scattering. It was shown that the solution could be improved considerably without doing more work than necessary for the first-order Magnus approximation, if the curve crossing was exploited within the framework of the first-order Magnus approximation. Since the modified Magnus approximation is straightforwardly applicable in the case of predissociation, we shall present only the important results with necessary definitions.

Since the major contribution to the transition amplitudes comes from the neighborhood of the curve-crossing point, in the modified Magnus approximation the solution of the equation of motion may be written in terms of $\mathbf{W}(0)$ at the crossing point and its correction $\mathbf{P}(v)$

$$\mathbf{X}(v) = \exp[\tfrac{1}{2}\,\mathbf{W}(0)(v - v_2)]\,\exp[\mathbf{P}(v)]\mathbf{X}_0 \quad , \tag{9.34}$$

where the variable v is such that

$$dv = (1 + u^2)^{-1}\,du \quad , \qquad -\pi/2 \le v \le \pi/2$$

and the correction factor $\mathbf{P}(v)$ is approximated by

$$\mathbf{P}(v) = \frac{1}{2} \int_{v_2}^{v} dv[\mathbf{W}(v) - \mathbf{W}(0)] + \frac{1}{4} \int_{v_2}^{v} dv\,\sin(v - v_2)[\mathbf{W},\mathbf{W}(0)]$$

$$+ \frac{1}{2} \int_{v_2}^{v} dv\,\sin^2 \frac{1}{2}(v - v_2)\,\mathbf{W}(0)[\mathbf{W}(0),\mathbf{W}] \quad , \tag{9.35}$$

which can easily be obtained by solving the equation of motion for $P(v)$ to the first-order Magnus approximation. The procedure is the same as that already discussed in Chap.7. Therefore, the matrix elements a_{ij} in (9.15-20) are given by

$$a_{ij} = \left\{ [\cos\tfrac{1}{2}(v_3 - v_2)\, \mathbf{I} + \sin\tfrac{1}{2}(v_3 - v_2)\, \mathbf{W}(0)]\exp[P(v_3)] \right\}_{ij} \qquad (9.36)$$

where v_2 and v_3 are the end points of the integration defined by

$$v_2 = \tan^{-1} u_2 \quad \text{and} \quad v_3 = \tan^{-1} u_3 \quad ,$$

with the meanings of u_2 and u_3 being the same as in the previous section.

If we apply the stationary-phase method as a further approximation method for the integrals in (9.35) as was done in Sects.7.2.2 and 9.1, the matrix $\mathbf{W}(0)$ takes the following remarkably simple form

$$\mathbf{W}(0) = D[\tau] \quad , \qquad (9.37a)$$

and the matrix $\mathbf{W}(v)$ the form

$$\mathbf{W}(v) = D[\tau + \psi^{(-)}]$$

$$= D[\tau]\cos\psi^{(-)} + D[\tau + \pi/2]\sin\psi^{(-)} \qquad (9.37b)$$

where the phases τ and $\psi^{(-)}$ are defined by (9.32c,d) respectively and the matrix $D[x]$ is defined in (7.65c).

It is useful for calculating P to note the following identities related to $D[\tau]$ and $D[\tau + \pi/2]$:

$$D^2[\tau] = -\mathbf{I} \quad ,$$

$$D[\tau]D[\tau + \pi/2] = \Sigma \quad ,$$

$$[D[\tau], D[\tau + \pi/2]] = 2\Sigma \quad ,$$

where $[,]$ is a commutator and

$$\Sigma = \begin{pmatrix} \sigma & 0 \\ 0 & \sigma \end{pmatrix} \quad , \qquad \sigma = \begin{pmatrix} 0 & 1 \\ -1 & 0 \end{pmatrix} \quad .$$

Note that the matrix σ is related to a Pauli matrix σ_y: $\sigma = i\sigma_y$.

Let us define the following integrals which are similar to the integrals (7.64b):

$$s_2 = \frac{1}{2} \int_{v_2}^{v_3} dv \, \cos v \, \sin\psi^{(-)} \qquad (9.38a)$$

$$s_3 = \frac{1}{2} \int_{v_2}^{v_3} dv \, \sin v \, \sin \psi^{(-)} \tag{9.38b}$$

and also the following symbols:

$$\lambda_2 = \left\{ \left[\frac{1}{2} (v_3 - v_2) - J \right]^2 + s_2^2 + s_3^2 \right\}^{1/2} \tag{9.39}$$

$$\zeta = \sin^{-1} \left[\left(s_2^2 + s_3^2 \right)^{1/2} / \lambda_2 \right] \tag{9.40}$$

$$\delta = \tan^{-1}(s_2/s_3) \quad . \tag{9.41}$$

Then, the stationary-phase approximation for $P(v_3)$ is

$$P(v_3) = -\lambda_2 \Big\{ \cos\zeta \, D[\tau] + \sin\zeta \, \cos(\delta + \pi/2 + v_2) D[\tau + \pi/2]$$

$$+ \sin\zeta \, \sin(\delta + \pi/2 + v_2) \Sigma \Big\} \quad . \tag{9.42}$$

Notice that it has the same structure as (7.92). It is given in terms of
four integrals τ, J, s_2, and s_3 which are simple quadratures. Since the matrix $P(v_3)$ is involutional,

$$P^2(v_3) = -\lambda_2^2 I \quad , \tag{9.43}$$

we have

$$\exp[P(v_3)] = \cos\lambda_2 \, I + \lambda_2^{-1} \sin\lambda_2 \, P(v_3) \quad . \tag{9.44}$$

Putting this into (9.36) and making use of the properties of the matrix $D[x]$ listed above, it is now easy to calculate a_{ij}:

$$a_{11} = a_{33} = F_1 \quad , \quad a_{13} = -a_{31} = F_2 \cos\tau + F_3 \sin\tau$$

$$a_{21} = -a_{43} = F_4 \quad , \quad a_{23} = a_{41} = -F_2 \sin\tau + F_3 \cos\tau \tag{9.45}$$

where

$$F_1 = \cos\frac{1}{2}(v_3 - v_2) \cos\lambda_2 + \sin\frac{1}{2}(v_3 - v_2) \sin\lambda_2 \cos\zeta$$

$$F_2 = \sin\frac{1}{2}(v_3 - v_2) \cos\lambda_2 + \cos\frac{1}{2}(v_3 - v_2) \sin\lambda_2 \cos\zeta$$

$$F_3 = \cos(\delta + \tfrac{1}{2}\pi + \tfrac{1}{2}v_2 + \tfrac{1}{2}v_3)\sin\lambda_2 \sin\zeta$$

$$F_4 = \sin(\delta + \tfrac{1}{2}\pi + \tfrac{1}{2}v_2 + \tfrac{1}{2}v_3)\sin\lambda_2 \sin\zeta \quad . \tag{9.46}$$

Since the major contributions to the integrals J, s_2, and s_3 come from the region of u equal to zero, namely, the vicinity of the curve-crossing point, it is possible to extend the range of integration to infinity in both directions or, equivalently, put $v_2 = -\frac{1}{2}\pi$ and $v_3 = \frac{1}{2}\pi$. This procedure is consistent with the stationary-phase method employed here. Then, it is easy to see that (9.46) reduces to

$$F_1 = \sin\lambda_2 \cos\zeta \quad ,$$

$$F_2 = \cos\lambda_2 \quad ,$$

$$F_3 = \sin\lambda_2 \cos(\delta + \frac{1}{2}\pi)\sin\zeta \quad ,$$

$$F_4 = \sin\lambda_2 \sin(\delta + \frac{1}{2}\pi)\sin\zeta \quad . \tag{9.47}$$

This may be used for calculating approximate transition amplitudes.

When (9.21,24,45,47) are substituted into (9.15-18), the improved transition probability amplitudes are explicitly obtained in terms of four basic integrals J, s_2, s_3, and τ in the modified Magnus approximation:

$$T_{11} + (2i)^{-1} = (2i)^{-1} \exp(2in_1) \frac{\cos(\beta + \tau) + h_1 \tan^2\lambda_2 \exp(i\tau)\cos\beta}{\cos(\beta + \tau) + h_1 \tan^2\lambda_2 \exp(-i\tau)\cos\beta} \quad , \tag{9.48a}$$

$$T_{12} = i^{\ell} \exp(-n_2 + in_1) \frac{h_2 \tan\lambda_2 \sin\tau}{\cos(\beta + \tau) + h_1 \tan^2\lambda_2 \exp(-i\tau)\cos\beta} \quad , \tag{9.48b}$$

where

$$h_1 = g_0^* g_1 / g_2^* g_3 \quad , \tag{9.49}$$

$$h_2 = \exp(i\beta)\{(g_2^*/g_0)\cos\beta + i[(g_3/g_1)(\cos(\tau + \beta)/\sin\tau)$$
$$- (g_2^*/g_0)\cos\beta \tan\tau]\} \quad , \tag{9.50}$$

with

$$g_0 = \cos\zeta + i \sin(\delta + \frac{1}{2}\pi)\sin\zeta \quad ,$$

$$g_1 = 1 + \sin(\delta + \frac{1}{2}\pi)\tan\zeta \tan\beta \quad , \tag{9.51}$$

$$g_2 = 1 + i \cos(\delta + \frac{1}{2}\pi)\tan\lambda_2 \sin\zeta$$

$$g_3 = 1 + \cos(\delta + \frac{1}{2}\pi)\tan\lambda_2 \sin\zeta \tan(\beta + \tau) \quad .$$

It is useful to note here that ζ is generally small so that $g_i, i = 0,1,2,3$, are almost equal to unity. In the limit of vanishing ζ the parameter λ_2 also tends to $\frac{1}{2}\pi - J$. Therefore, we see that T_{11} and T_{12} given by (9.48a,b) in the modified Magnus approximation are reduced to those in (9.33a,b), respectively, in the limit.

9.3 A Multistate Model for Predissociation

In Chap.8 we have considered a multistate model for inelastic scattering where n-1 repulsive potential curves cross an attractive potential. The model is amenable to semianalytic approximation methods which furnish us with some insightful results for transition probability amplitudes. Here we adopt the same model of potentials to study multistate predissociation processes. We have seen that predissociation phenomena require equations of motion with mathematical structure similar to the corresponding inelastic scattering processes. In fact, it is possible to show that if one approaches the problem in the uniform semiclassical method, then the equation of motion (8.1) applies if the wave functions for the nth adiabatic potential are replaced by the semiclassical wave functions for the two-turning-point problem presented in Sect.9.1.

To use the results obtained in Chap.8, let us first discuss the three-state problem as depicted in Fig.8.1, which is the predissociation counterpart of the inelastic scattering process already considered, and then generalize the results to the n-state problem by means of rules similar to those in Chap.8. Therefore, this part of the discussion is completely parallel to the discussion in Sect.8.1 and consequently we shall be relying heavily on Chap.8, presenting only the equations and results relevant to predissociation.

For the n-state predissociation arising from the model depicted in Fig. 8.2, the transition probability amplitudes are formally given as follows:

$$T_{n-1,k} + (2i)^{-1}\delta_{n-1,k} = (2i)^{-1} \exp[i(\eta_{n-1} + \eta_k)]$$

$$\times \sum_{j=1}^{n} (a_{2k-1,j} + ia_{2k,j})\Delta_{n-1,j}/\Delta \quad ,$$

$$(k = 1,2,\ldots,n-1) \qquad (9.52a)$$

and

$$T_{n-1,n} = i^{\ell-1} \exp(i\beta_n) \exp(i\eta_{n-1} - \bar{\eta}_n)$$

$$\times \sum_{j=1}^{n} (a_{2n-1,j} \cos\beta_n + ia_{2n,j} \sin\beta_n)\Delta_{n-1,j}/\Delta \quad , \qquad (9.52b)$$

where Δ is an $n \times n$ determinant:

$$\Delta = \det|d_{ij}| \qquad (9.53)$$

with its elements defined by

$$d_{ij} = a_{2i-1,2j-1} - ia_{2i,2j-1} \quad , \quad (i = 1,2,\ldots,n-1 \ ; \quad j = 1,2,\ldots,n) \quad (9.54)$$

$$d_{nj} = a_{2n-1,2j-1} \cos\beta_n - a_{2n-2j-1} \sin\beta_n \quad , \quad (j = 1,2,\ldots,n)$$

$$\eta_j = \frac{1}{2} (\ell + \frac{1}{2})\pi - k_j r_j + \int_{r_j}^{\infty} (\lambda\Lambda_j - k_j) \quad , \quad (j = 1,2,\ldots,n-1)$$

$$\bar{\eta}_n = -|k_n|r_n + \int_{r_n}^{\infty} (\lambda|\Lambda_n| - |k_n|)$$

$$\beta_n = \lambda \int_{r_n}^{r_{n+1}} \Lambda_n \, dx \qquad (9.55)$$

and $\Delta_{n-1,j}$ is the $(n-1,j)$ cofactor of the determinant. Here Λ_n is the nth adiabatic momentum and the turning points are defined in Fig.8.2.

In the pairwise curve-crossing approximation taken for the corrresponding inelastic scattering problem, the solution of the equation of motion was given as a product of sequential two-state solutions. Therefore, for the three-state model represented by Fig.8.1 the approximate solution of the equation of motion is in the same form as (8.11) where the only change required is in the uniform semiclassical wave functions A_3 and B_3 appearing in the matrices W_1 and W_2. Therefore, in the first Magnus approximation the matrix elements a_{ij} are again given by

$$a_{ij} = [\exp(f_2)\exp(f_1)]_{ij} \ . \qquad (9.56)$$

If the energy is sufficiently high, then the stationary-phase method is reasonable. When this method is applied for evaluating the integrals involved in calculating a_{ij}, it is clear in view of the same structure of the matrices W_1 and W_2 and similar behavior of the functions involved, that (8.20a) will be recovered for the matrices along with the same attendant equations; see (7.20b-21b). The results of a_{ij} so calculated are summarized in Table 9.1. Using Table 9.1 to calculate the transition amplitudes (9.52a, b), we obtain them for $n = 3$ in the first-order Magnus stationary-phase approximation:

$$T_{22} + (2i)^{-1} = (2i)^{-1} \exp(2i\eta_2)[\cos\phi_1 + \chi_1 \exp(-i\tau_1)\cos\phi_2$$

Table 9.1. Relevant matrix elements a_{ij} in the stationary-phase approximation[a]

j	1	3	5
a_{1j}	c_1	$s_1 \cos\tau_1$	0
a_{2j}	0	$-s_1 \sin\tau_1$	0
a_{3j}	$-s_1 c_2 \cos\tau_1$	$c_1 c_2$	$s_2 \cos\tau_2$
a_{4j}	$-s_1 c_2 \sin\tau_1$	0	$-s_2 \sin\tau_2$
a_{5j}	$s_1 s_2 \cos(\tau_1 + \tau_2)$	$-c_1 s_2 \cos\tau_2$	c_2
a_{6j}	$s_1 s_2 \sin(\tau_1 + \tau_2)$	$-c_1 s_2 \sin\tau_2$	0

[a] $c_i = \cos J_i$; $s_i = \sin J_i$ (i = 1,2).

$$+ \; x_1 x_2 \exp(-i\tau_{12}^{+})\cos\beta_3 + x_2 \exp(i\tau_{12}^{-})\cos\beta_3]\Delta^{-1} \tag{9.57a}$$

$$T_{21} = \exp[i(\eta_1 + \eta_2)][x_1 x_2(x_2 + 1)]^{1/2} \cos\beta_3 \; \sin\tau_1 \; \exp(-i\tau_{12}^{+})\Delta^{-1} \tag{9.57b}$$

$$T_{23} = i^{\ell-1} \exp[i\eta_2 - \eta_3]\left\{ x_2^{1/2} \cos\phi_1 \exp(-i\tau_2) + x_1 \exp(-i\tau_{12}^{+})\cos\phi_2 \right.$$
$$\left. - \; x_2^{1/2}[\exp(i\tau_1) + x_1 x_2^{1/2} \exp(-i\tau_{12}^{+})]\cos\beta_3 \right\}\Delta^{-1} \tag{9.57c}$$

where

$$\Delta = \cos\phi_1 + x_1 \exp(-i\tau_1)\cos\phi_2 + x_2(1 + x_1)\exp(-i\tau_{12}^{+})\cos\beta_3$$

$$\phi_1 = \beta_3 + \tau_1 + \tau_2$$

$$\phi_2 = \beta_3 + \tau_2 \tag{9.58}$$

$$\tau_{12}^{\pm} = \tau_1 \pm \tau_2$$

$$x_i = \cot^2 J_i \quad (i = 1,2) \; . \tag{9.59}$$

Here T_{23} is essentially the amplitude for the resonance state supported by the predissociative potential V_2 and therefore is not directly measurable in the laboratory. Equation (9.57a) represents the transition amplitude for elastic scattering of state 1 subject to the predissociative potential V_3 and the repulsive potentials V_1 and V_2.

If the coupling potential V_{12} is equal to zero everywhere, then the parameter x_1 is equal to zero and the phase τ_1 must be taken equal to zero. In this case, T_{21} vanishes and T_{22} reduces to the transition amplitude of a two-state problem with potentials V_2 and V_3 and coupling potential V_{23}:

$$T_{22} + (2i)^{-1} = \frac{\exp(2i\eta_2)}{2i} \frac{\cos\phi_2 + x_2 \exp(i\tau_2)\cos\beta_3}{\cos\phi_2 + x_2 \exp(-i\tau_2)\cos\beta_3} \quad . \tag{9.60}$$

The phase integrals appearing in the transition amplitudes are amenable to the following interpretation:

$$\tau_{12} + \tau_{23} + \beta_3 = \lambda \int_{r_1}^{r_4} (E - V')^{1/2} dr \tag{9.61}$$

$$\tau_{23} + \beta_3 = \lambda \int_{r_2}^{r_4} (E - V'')^{1/2} dr \quad , \tag{9.62}$$

where

$$V' = \zeta(r - r_{c1})V_{a1} + [1 - \zeta(r - r_{c1})]V_{a2}$$

$$\quad + [1 - \zeta(r - r_{c2})]V_{a3}$$

$$V'' = \zeta(r - r_{c2})V_{a2} + [1 - \zeta(r - r_{c2})]V_{a3}$$

$$\zeta(r - r_{ci}) = \begin{cases} 1 & r < r_{ci} \\ 0 & r > r_{ci} \end{cases} \quad .$$

As the coupling strengths of V_{12} and V_{23} decrease, the potential V' tends to V_3, while the potential V'' approaches the upper adiabatic potential made up with V_1 and V_3 (Fig.8.1). Therefore, (9.61) can be associated with the diabatic vibrational states of V_3, while (9.62) and β_3 may be associated with the adiabatic vibrational states.

The transition factor x_i is equivalent to the LZS factor $[\exp(2s) - 1]$ of (7.38,37) associated with the curve-crossing point i. If x_i is replaced with the LZS factor above, the transition amplitudes become similar to those obtained by *Sink* and *Bandrauk* [9.9], although there are some differences in the phase factors and in the way the transition factors appear in the numerators. In fact, their method of constructing the S matrix is difficult to follow. We note that *Dubrovskii* [8.10] has also obtained similar results by using LZS-type theory.

The three-state model considered above suggests a generalization to an n-state problem depicted in Fig.8.2, which is the predissociation counterpart of the n-state problem considered in Chap.8. The equation of motion has the same form as (8.1) and can be solved by the pairwise curve-crossing approximation. In that case the procedure of solution is completely parallel to that in Chap.8. Again, the construction of the transition probability amplitudes can be accomplished by suitably modifying the set of rules stated in Chap.8 for constructing the determinant $\Delta = \det|d_{ij}|$ defined in (9.53).

Let us define $c_0 = 1$ and $c_n = 1$. The rules are as follows.

Rule 1a. The jth element d_{ji} of the ith column, where $i \leq j \leq n - 1$, is given by

$$d_{ji} = c_{i-1}c_j \prod_{k=i}^{j-1} [-s_k \exp(-i\tau_k)] \quad , \quad (i = 1,2,\ldots,n) \quad ; \tag{9.63}$$

the product must be put equal to 1 when the upper limit is less than the lower limit.

Rule 1b. The element d_{ni} is given by

$$d_{ni} = c_{i-1}c_n \prod_{k=1}^{n-1} (-s_k)\cos\left(\beta_n + \sum_{k=1}^{n-1} \tau_k\right) \quad . \tag{9.64}$$

Rule 2. For $j < i$

$$d_{i-1,i} = s_{i-1} \exp(i\tau_{i-1}) \quad (i = 1,2,\ldots,n)$$

$$d_{ji} = 0 \quad \text{for all} \quad j < i - 1 \quad . \tag{9.65}$$

This rule defines the upper triangle of the determinant, while Rules 1a and 1b define the lower triangle including the main diagonal. The following rule is for constructing $T_{n-1,k}$. We construct an auxiliary determinant $\Delta^{(*)}$ from the determinant given by the above rules.

Rule 3. $\quad \Delta^{(*)} = \det|d_{ji}^{(*)}| \quad , \tag{9.66}$

where

$$d_{ji}^{(*)} = d_{ji}^*$$

for $1 \leq j \leq n - 1$ and $1 \leq i \leq n$,

$$d_{ni}^{(*)} = c_{i-1}c_n \prod_{k=1}^{n-1} (-s_k)$$

$$\times \left[\cos\left(\sum_{k=i}^{n-1} \tau_k\right)\cos\beta_n + i \sin\left(\sum_{k=i}^{n-1} \tau_k\right)\sin\beta_n \right] \quad , \tag{9.67}$$

$$(i = 1,2,\ldots,n)$$

and

$$d_{nn}^{(*)} = c_{n-1}c_n \cos\beta_n \quad . \tag{9.68}$$

The transition amplitudes are then given by

$$T_{n-1,j} + (2i)^{-1}\delta_{j,n-1} = (2i)^{-1} \exp[i(n_{n-1} + n_j)]$$

$$\times \sum_{k=1}^{n} \Delta_{n-1,k}d_{jk}^{(*)}/\Delta \tag{9.69a}$$

and

$$T_{n-1,n} = i^{\ell-1} \exp(i\beta_n)\exp(in_{n-1} - \bar{n}_n)$$

$$\times \sum_{k=1}^{n} \Delta_{n-1,k}d_{nk}^{(*)}/\Delta \quad . \tag{9.69b}$$

These are the desired results for the predissociative n-curve-crossing problem in the first-order Magnus stationary-phase approximation. These rules can be shown to be valid by induction and verified for lower n's by explicit calculation. We show the determinant for n = 4 as an example below:

$$\Delta = \begin{vmatrix} c_1 & s_1 \exp(i\tau_1) & 0 & 0 \\ -c_2s_1 \exp(-i\tau_1) & c_1c_2 & s_2 \exp(i\tau_2) & 0 \\ c_3s_1s_2 \exp(-i\tau_{12}) & -c_1c_3s_2 \exp(-i\tau_2) & c_2c_3 & s_3 \exp(i\tau_3) \\ -s_1s_2s_3 \cos(\beta_3+\tau_{123}) & c_1s_2s_3 \cos(\beta_3+\tau_{23}) & -c_2s_3 \cos(\beta_3+\tau_3) & c_3 \cos\beta_3 \end{vmatrix}$$

where $\tau_{12...\ell}$ is defined by (8.32).

The determinant may be more explicitly calculated in the form

$$\Delta = \left[(-)^{n+1} \prod_{k=1}^{n-1} (-s_k^2)\exp(i\tau_k) \right] \sum_{\ell=1}^{n} \xi_\ell \cos(\beta_n + \tau_{\ell,\ell+1,...,n-1}) \tag{9.70}$$

where

$$\xi_\ell = \Delta_{n\ell}c_{\ell-1}/\Delta_{n1} \prod_{k}^{\ell-1} (-s_k) \quad .$$

This may be recast into

$$\xi_\ell = x_{\ell-1} \prod_{k=1}^{\ell-2} (1 + x_k)\exp(-i\tau_{12\ldots\ell-1}) \quad ,$$

where x_i is defined by (9.59) and the product must be put equal to zero if the upper limit is less than the lower limit.

With this form of Δ, we are now able to determine the energy levels from the zeros of the equation,

$$\sum_{\ell=1}^{n} \xi_\ell \cos(\beta_n + \tau_{\ell,\ell+1,\ldots,n-1}) = 0 \quad . \tag{9.71}$$

Let us separate ξ_ℓ into real and imaginary parts:

$$\xi_\ell = \xi_\ell^{(R)} + i\xi_\ell^{(I)}$$

and assume that an energy E_L exists such that

$$\beta_n + \tau_{1,2,\ldots,n-1} = (L + 1/2)\pi \quad ,$$

where $L = 0,1,2,\ldots$. Then in the vicinity of E_L we may write (9.71) in the form

$$(-)^{L+1}\pi(\partial L/\partial E_L)(E - E_L - \Delta E_L - i\Gamma_L/2) = 0 \quad ,$$

where the level shift ΔE_L and the level width Γ_L are given by

$$\Delta E_L = \sum_{\ell=2}^{n} \Delta E_L^{(\ell)} = (-)^L \sum_{\ell=2}^{n} \xi_\ell^{(R)}(E_L)\cos(\beta_n + \tau_{\ell,\ell+1,\ldots,n-1})[\pi(\partial L/\partial E_L)]^{-1} \tag{9.72}$$

$$\Gamma_L = \sum_{\ell=2}^{n} \Gamma_L^{(\ell)} = (-)^L \sum_{\ell=2}^{n} \xi_\ell^{(I)}(E_L)\cos(\beta_n + \tau_{\ell,\ell+1,\ldots,n-1})[\pi(\partial L/\partial E_L)]^{-1} \quad . \tag{9.73}$$

It can be shown that

$$\xi_\ell^{(R)} = x_{\ell-1} \prod_{k=1}^{\ell-2} (1 + x_k)\cos\tau_{12\ldots\ell-1} \tag{9.74}$$

$$\xi_\ell^{(I)} = x_{\ell-1} \prod_{k=1}^{\ell-2} (1 + x_k)\sin\tau_{12\ldots\ell-1} \quad .$$

Using these relations, we finally obtain

$$\Gamma_L^{(\ell)} = \chi_{\ell-1}(E_L) \prod_{k=1}^{\ell-2} [1 + \chi_k(E_L)]\sin^2 \tau_{12\ldots\ell-1}[\pi(\partial L/\partial E_L)]^{-1} \qquad (9.75)$$

$$\Delta E_L^{(\ell)} = \frac{1}{2} \chi_{\ell-1}(E_L) \prod_{k=1}^{\ell-2} [1 + \chi_k(E_L)]\sin 2\tau_{12\ldots\ell-1}[\pi(\partial L/\partial E_L)]^{-1} \quad .$$

Notice that there are $(\ell-1)$ transition factors $\chi_k(E_L)$ appearing in the ℓth partial width $\Gamma_L^{(\ell)}$ and the total width is a sum of partial widths associated cumulatively with the curve-crossing points, but the partial widths are not in one to one correspondence with the crossing points. The additivity of partial widths is reminiscent of the perturbation theory result. However, it should be emphasized that the perturbation result does not have the cumulative effect on $\Gamma_L^{(\ell)}$ of the crossing points that exists in (9.73).

10. A Multisurface Scattering Theory

The curve-crossing problems studied in the previous two chapters are one-dimensional cases of more general multisurface scattering. From the chemist's viewpoint multisurface scattering is of considerable interest, because there are many important chemical reactions that involve more than one potential surface. Such reactions were first studied theoretically by *Tully* and *Preston* [10.1] and later studies include the Feynman path integral method by *George* and collaborators [10.2,3] and others [10.4,5]. The Feynman path integral method contains an attractive feature that appears well suited for studying electronic transitions involving more than one potential surface. Therefore, it is logical that the complex trajectory method is based on this approach. The complex trajectory method as formulated by *George* and co-workers blends the Schrödinger picture for electronic motion, the Feynman picture for nuclear motion, as well as the WKB approximation for the electronic and nuclear transition probabilities. The WKB approximation for the electronic transition probabilities turns out to be basically a slight generalizatior of the well-known result by *Zener* [7.1] in the sense that the phase integrals are evaluated numerically by solving the classical equations of motion in the complex time plane. However, we recall that Zener's theory treats not electronic motion per se, but nuclear motion subject to the effective potentials provided by the electrons in the Born-Oppenheimer approximation. Therefore, the appearance of the LZS factor in the Feynman path integral for nuclear motion is rather curious. When the second-order differential equation in time in the Zener theory is solved and appropriate boundary conditions are taken into account, transition probabilities with approximate phases are obtained for nuclear motions and the problem is essentially solved [4.39,9.9]. This in effect obviates the step of evaluating the path integrals needed in the Miller-George theory.

In this chapter we examine the multisurface scattering problem from a different perspective, hoping to understand the complex trajectory method. The subsequent discussion is based on an unpublished paper (1976) by the author,

quite similar in content to the recent theory by *Bjorken* and *Orbach* [10.6].
Some formal theoretical aspects are a little more developed than theirs, although further studies are still required in the future. It is presented
here as a suggestion of an area of semiclassical theory that requires further
studies. The point to be made here is that there is an equation for transition probabilities for nuclear motion that can be solved by computing the
classical trajectories for particles (nuclei) moving on multidimensional
potential surfaces.

10.1 Multisurface WKB Solutions

Imagine a system of atoms which move in potential fields generated by electrons in the diabatic representation. Then the electronic energy will be the
potential energy surfaces and the Hamiltonian may be generally written in a
matrix form

$$
H = \begin{pmatrix}
H_1 & V_{12} & \cdots & V_{1n} \\
V_{21} & H_2 & \cdots & V_{2n} \\
\vdots & \vdots & & \vdots \\
V_{n1} & V_{n2} & \cdots & H_n
\end{pmatrix}
\tag{10.1}
$$

where

$$
H_i = - \sum_{\ell=1}^{f} (\hbar^2/2m_\ell)(\partial^2/\partial r_\ell^2) + V_i(r_1,\ldots,r_f)
$$

$$
V_{ij} = V_{ij}(r_1,\ldots,r_f) \quad ,
\tag{10.2}
$$

with r_j denoting the coordinate vector of atom j. The subscripts i,j attached
to H_i and V_{ij} denote the diabatic states. Further, V_{ij} represents the coupling between the states i and j. The Born-Oppenheimer approximation is implicit for the nuclear motion.

The Schrödinger equation is

$$
i\hbar(\partial/\partial t)\Psi = H\Psi(r,t) \quad .
\tag{10.3}
$$

We look for WKB wave functions as approximate solutions for (10.3). To this
purpose, we write Ψ in the form

$$\Psi = \begin{pmatrix} A_1 \\ A_2 \\ \cdot \\ \vdots \\ A_n \end{pmatrix} \exp(i\lambda S) \qquad (\lambda = \hbar^{-1}) \quad , \tag{10.4}$$

where the phase S and the amplitudes A_i are functions of $r_1, r_2, \ldots, r_n$, and t. The form of the wave function is a generalization of $T(p|p_0)$ in (6.17,18). Therefore, it is easily seen that a similar method of solution will apply in the present case as well. Indeed, only a slight generalization will be necessary to obtain the solution.

Substitution of (10.4) into (10.3) yields

$$i\hbar(\partial/\partial t)\mathbf{A} - \mathbf{A}(\partial S/\partial t) = H[\{p_\ell + (\partial/\partial r_\ell)S\}, \{r_\ell\}]\mathbf{A} \tag{10.5}$$

where

$$p_\ell = -i\hbar(\partial/\partial r_\ell)$$

quantum mechanically. The column vector in (10.4) is abbreviated by $\mathbf{A}$:

$$\mathbf{A} = \begin{pmatrix} A_1 \\ A_2 \\ \vdots \\ \vdots \\ A_n \end{pmatrix} \quad . \tag{10.6}$$

Note that (10.5) is a matrix equation unlike its counterpart in Chap.6. Since the matrix H in (10.5) is generally complex, we write it in terms of real and imaginary parts:

$$H = H_R + i H_{im} \quad . \tag{10.7}$$

Since we choose A_i and S to be real, (10.5) may be written in two equations for real and imaginary parts:

$$[H_R(\{p_\ell + (\partial/\partial r_\ell)S\}, \{r_\ell\}) + (\partial S/\partial t)]\mathbf{A} = 0 \tag{10.8}$$

$$\hbar(\partial A/\partial t) = H_{im}(\{p_\ell + (\partial S/\partial r_\ell)\}, \{r_\ell\})\mathbf{A} \quad . \tag{10.9}$$

If the time-independent Schrödinger equation is used,

$$H\Psi = E\Psi \quad , \tag{10.10}$$

then the equations corresponding to (10.8,9) are

$$[H_R(\{p_\ell + (\partial S/\partial r_\ell)\}, \{r_\ell\}) - EI]\mathbf{A} = 0 \quad , \tag{10.8a}$$

$$H_{im}(\{p_\ell + (\partial S/\partial r_\ell)\}, \{r_\ell\})\mathbf{A} = 0 \quad . \tag{10.9a}$$

167

By explicitly performing the calculation in the coordinate representation for H, we find

$$H_R = \left\{ \sum_j [(2m_j)^{-1}(\partial S/\partial r_j)^2 - \hbar^2(2m_j)^{-1}(\partial^2 S/\partial r_j^2)] + (\partial S/\partial t) \right\} I + V \qquad (10.11)$$

$$H_{im} = -\hbar \sum_j [(2m_j)^{-1}(\partial^2 S/\partial r_j^2) + m_j^{-1}(\partial S/\partial r_j)\cdot(\partial/\partial r_j)]I , \qquad (10.12)$$

where V is the potential matrix

$$V = \begin{pmatrix} V_1 & V_{12} & \cdots & V_{1n} \\ V_{21} & V_2 & \cdots & V_{2n} \\ \vdots & \vdots & & \vdots \\ V_{n1} & V_{n2} & \cdots & V_n \end{pmatrix} . \qquad (10.13)$$

Equations (10.8,9) may be solved by an asymptotic expansion method similar to that in Chap.6. We write

$$S = S^{(0)} + \lambda^{-1}S^{(1)} + \lambda^{-2}S^{(2)} + \dots \qquad (10.14)$$

$$A = A^{(0)} + \lambda^{-1}A^{(1)} + \lambda^{-2}A^{(2)} + \dots , \qquad (10.15)$$

and then, substituting these into (10.8,9), obtain a hierarchy of equations for $S^{(a)}$ and $A^{(a)}$ (a = 0,1,2,...), led by the pair of equations

$$\left\{ \left[\sum_j (2m_j)^{-1}(\partial S^{(0)}/\partial r_j)^2 + (\partial S^{(0)}/\partial t) \right] I + V \right\} A^{(0)} = 0 \qquad (10.16)$$

$$- \sum_j \left[(2m_j)^{-1}(\partial^2 S^{(0)}/\partial r_j^2) + m_j^{-1}(\partial S^{(0)}/\partial r_j)\cdot(\partial/\partial r_j) \right] A^{(0)} = (\partial A^{(0)}/\partial t) . \qquad (10.17)$$

Since we do not consider the terms of order λ^{-1} or higher, they are not written here.

Note that (10.16,17) are matrix equations while the corresponding equations in Chap.6 are not. To put them, especially (10.16), into forms similar to those in Chap.6, we introduce a linear (orthogonal) transformation U such that it diagonalizes V:

$$A^{(0)} = UZ \qquad (10.18)$$

and

$$W = U^{-1}VU = \begin{pmatrix} W_1 & & & 0 \\ & W_2 & & \\ & & \ddots & \\ 0 & & & W_n \end{pmatrix} . \qquad (10.19)$$

On such a linear transformation, there follows from (10.16) a set of equations

$$\left[\sum_j (2m_j)^{-1}(\partial S_\alpha^{(0)}/\partial r_j)^2 + (\partial S_\alpha^{(0)}/\partial t) + W_\alpha\right] Z_\alpha = 0 \qquad (10.20)$$

for all $\alpha = 1, 2, \ldots, n$. Here the subscript α to $S^{(0)}$ denotes that the latter will be different for different α. The diagonal components W_α in (10.19) are simply the adiabatic potential surfaces for the system. Since Z_α are not equal to zero in general, we see that $S_\alpha^{(0)}$ obeys the equation

$$\sum_j (2m_j)^{-1}(\partial S_\alpha^{(0)}/\partial r_j)^2 + (\partial S_\alpha^{(0)}/\partial t) + W_\alpha = 0 \quad , \qquad (10.21)$$

which is the Hamilton-Jacobi equation for motion of atoms on the adiabatic surface W_α.

The solution of the equation is an action integral

$$S_\alpha^{(0)} = \int_{t_1}^{t_2} dt \, L_\alpha(\mathbf{r},\dot{\mathbf{r}}) \quad , \qquad (10.22)$$

where L_α is the Lagrangian corresponding to motion on the αth adiabatic surface,

$$L_\alpha = \sum_j \mathbf{p}_j \cdot \dot{\mathbf{r}}_j - W_\alpha \quad . \qquad (10.23)$$

When the transformation (10.18) is substituted into (10.17), we obtain the differential equation for Z:

$$\frac{\partial Z}{\partial t} + \sum_j m_j^{-1}\left[\frac{1}{2}\frac{\partial^2 S^{(0)}}{\partial r_j^2} + \frac{\partial S^{(0)}}{\partial r_j} \cdot \frac{\partial}{\partial r_j}\right] Z = MZ \quad , \qquad (10.24)$$

$$M = U^{-1}\left[(\partial U/\partial t) + \sum_j m_j^{-1}(\partial S^{(0)}/\partial r_j) \cdot (\partial U/\partial r_j)\right] = -U^{-1}(dU/dt) \quad . \qquad (10.25a)$$

Note that we regard the quantities in square brackets in (10.24,25a) as diagonal matrices. For the second equality in (10.25a) we have used

$$\mathbf{p}_j = (\partial S^{(0)}/\partial r_j) \quad . \qquad (10.25b)$$

Since the transformation U is orthogonal, the diagonals of M are equal to zero:

$$M_{jj} = \left(U^{-1}\frac{d}{dt}U\right)_{jj} = 0 \quad .$$

Therefore, if we neglect the off-diagonal components of M in (10.24), we obtain the approximate solution $Z^{(0)}$ from

$$\frac{\partial Z^{(0)}}{\partial t} + \sum_j m_j^{-1} \left[\frac{1}{2} \frac{\partial^2 S^{(0)}}{\partial r_j^2} + \frac{\partial S^{(0)}}{\partial r_j} \frac{\partial}{\partial r_j} \right] Z^{(0)} = 0 \quad , \tag{10.24a}$$

which together with (10.22) gives the WKB solution. Recall that the solution of this equation was discussed in Chap.6. It is given in terms of the Jacobian of transformation:

$$Z_\alpha^{(0)} = \Delta_\alpha^{1/2} = \left(\frac{\partial(r_1,\ldots,r_f)}{\partial(p_1^0,\ldots,p_f^0)} \right)^{1/2} \quad , \quad (\alpha = 1,2,\ldots,n) \quad , \tag{10.26}$$

where p_i^0 is the momentum at t_2, if r_i is evaluated at t_1.

By collecting these results, we obtain the WKB solution to (10.3):

$$\Psi(t) = U\left\{ \Delta_1^{1/2} \exp(i\lambda S_1^{(0)}), \ \Delta_2^{1/2} \exp(i\lambda S_2^{(0)}), \ \ldots, \ \Delta_n^{1/2} \exp(i\lambda S_n^{(0)}) \right\} \quad , \tag{10.27}$$

where the curly brackets stand for a column vector. This solution must be corrected by including the neglected off-diagonal elements of M in (10.24). To indicate how one might proceed about it, we recast (10.24):

$$\frac{d}{dt} Z = \left[M - \sum_j \frac{1}{2} \frac{\partial}{\partial r_j} (dr_j/dt) \right] Z \equiv KZ \tag{10.24b}$$

for which we have used (10.25b). The solution may be written in terms of a product integral

$$Z(t) = \prod_{t_0}^{t} \exp[K(s)ds]Z_0 \quad ,$$

where Z_0 is the initial value of Z. This integral may be evaluated by computing the classical trajectories for the system. However, since the wave functions so corrected do not generally produce inelastic transition probabilities the same reason why the wave functions in Sect.5.1 do not yield inelastic transition probabilities — unless suitable mixing schemes are introduced for different adiabatic channels, we shall not pursue solution of (10.24b) here.

Let us now return to (10.21) and write

$$S_\alpha^{(0)} = F_\alpha - Et \quad .$$

Then the time-independent Hamilton-Jacobi equation follows:

$$\sum_j (2m_j)^{-1} (\partial F_\alpha / \partial r_j)^2 + W_\alpha = E \quad . \tag{10.28}$$

This equation implies that there are two distinct solutions, since $\pm F_\alpha$ both satisfy the same equation. To make this more explicit, we define k_α by the two solutions,

$$F_\alpha = \pm k_\alpha$$

and then write the wave functions in the form

$$\Psi^{(\pm)} \equiv \left\{\Psi_1^{(\pm)}, \ldots, \Psi_n^{(\pm)}\right\}$$

$$= U\left\{\Delta_1^{1/2} \exp(-i\lambda Et \pm i\lambda k_1), \ldots, \Delta_n^{1/2} \exp(-i\lambda Et \pm i\lambda k_n)\right\} \quad . \qquad (10.28a)$$

The solution of (10.20,17) or the WKB solution for (10.3), even if (10.24b) were solved exactly, would not be complete as yet, since (10.28a) must be made to satisfy appropriate boundary conditions. Although we are not going to solve explicitly the boundary value problems which cannot be achieved in generality, we must provide the formalism with some formal solutions of them. Since the subsequent formalism is time dependent, we shall consider (10.28) again in connection with the boundary conditions.

Since the Hamilton-Jacobi equation (10.28) separates into the translational and internal parts as the collision partners are at a large distance from each other, and in that state some of the internal parts separate even further, we group the coordinates into subgroups into which the system separates asymptotically at large relative distances of the center of gravity of the collision partners. Thus

$$(r_1, r_2, \ldots, r_f) = (Q_{\alpha 0}, Q_{\alpha 1}, Q_{\alpha 2}, \ldots, Q_{\alpha s})$$

$$\equiv (Q_\alpha) \quad . \qquad (10.29)$$

Here $Q_{\alpha 0}$ stands for the relative distance between the centers of mass of the collision partners, and $Q_{\alpha i}$ ($i \neq 0$), the internal coordinates of the separated partners. Since the regrouping (10.29) is canonical, the Hamilton-Jacobi equation stays invariant:

$$\sum_j (2\mu_j)^{-1} (\partial F_\alpha / \partial Q_{\alpha j})^2 + W_\alpha(Q_{\alpha 0}, \ldots, Q_{\alpha n}) = E \quad , \qquad (10.30)$$

where μ_j are the masses in the new coordinate system. Now let us define $W_{\alpha \ell}(Q_{\alpha \ell})$ such that

$$\lim_{|R_\alpha| \to \infty} W(Q_{\alpha 0}, Q_{\alpha 1}, \ldots, Q_{\alpha s}) = \sum_{\ell=1}^{s} W_{\alpha \ell}(Q_{\alpha \ell}) \quad . \qquad (10.31)$$

Then in the same limit the translational part of F_α separates from the internal part:

$$\lim_{|Q_{\alpha 0}| \to \infty} F_\alpha = F_\alpha^{(int)} + F_{\alpha 0} \quad , \qquad (10.32)$$

171

where $F_\alpha^{(int)}$ satisfies the internal Hamilton-Jacobi equation,

$$\sum_{j \in \ell} (2\mu_j)^{-1} (\partial F_\alpha^{(int)}/\partial Q_{\alpha j})^2 + W_{\alpha \ell}(Q_{\alpha \ell}) = E_\ell^{(\alpha)} \quad , \tag{10.33}$$

where $\ell = 1,2,\ldots,s$, and $F_{\alpha 0}$ also satisfies the Hamilton-Jacobi equation

$$\sum_{j \in 0} (2\mu_j)^{-1} (\partial F_{\alpha 0}/\partial Q_{\alpha j})^2 = E_0^{(\alpha)} \quad . \tag{10.34}$$

Here the energies are such that

$$E = E_0^{(\alpha)} + \sum_{\ell=1}^{s} E_\ell^{(\alpha)} \quad .$$

The solutions of the internal Hamilton-Jacobi equations (10.33) furnish us with the WKB solutions for the internal degrees of freedom. When the boundary conditions are appropriately imposed, they yield the quantization conditions for the eigenvalues and the corresponding wave functions for internal mo-tions undergoing periodic motions. Under the transformation (10.29) the am-plitudes $A^{(0)}$ may be written in terms of new variables and the corresponding momenta at $t = t_1$:

$$A^{(0)} = U\left\{\Delta_1^{1/2}, \ldots, \Delta_n^{1/2}\right\} \quad ,$$

where

$$\Delta_\alpha = \frac{\partial (Q_{\alpha 0}, Q_{\alpha 1}, \ldots, Q_{\alpha s})}{\partial (P_{\alpha 0}^0, P_{\alpha 1}^0, \ldots, P_{\alpha s}^0)} \quad .$$

The Jacobian will factorize into translational and internal Jacobians as the relative distance between collision partners increases, i.e., as $|Q_{\alpha 0}| \to \infty$. Here for our purpose it is sufficient to assume that such WKB wave functions denoted by $\psi_n^{(\alpha)}$ are already obtained. We assume that they form a complete set.

We then define projection operators in terms of them:

$$P_n^{(\alpha)} = |\psi_n^{(\alpha)}\rangle\langle\psi_n^{(\alpha)}| \quad . \tag{10.35}$$

The WKB wave functions for relative motion are then defined by the projections of $\psi_\alpha^{(\pm)} = \Delta_\alpha^{1/2} \exp(\pm ik_\alpha)$:

$$u_n^{(\pm)} = P_n^{(\alpha)}\psi_\alpha^{(\pm)} \quad . \tag{10.36}$$

Therefore, we may write

$$\psi_\alpha^{(\pm)} = \sum_n u_{\alpha n}^{(\pm)} \psi_n^{(\alpha)} \quad , \tag{10.37}$$

which motivates the choice of the following basis set:

$$\{\psi_{\alpha n}^{(\pm)} \, , \quad n = 1,2,\dots\} \equiv \{u_{\alpha n}^{(\pm)} \psi_n^{(\alpha)} \; ; \quad n = 1,2,\dots\} \tag{10.38}$$

for all α. Here the projection $u_{\alpha n}^{(\pm)}$ must also satisfy suitable scattering boundary conditions. Note, however, that $u_{\alpha n}^{(\pm)}$ are the WKB solutions for translational motion in the adiabatic representation and do not yield the inelastic translation probabilities unless a further correction is made to them. For this we proceed in the same manner as in the previous chapters. However, it must be remarked that one can integrate the next members of the hierarchy following the set (10.16,17) to obtain the desired correction, although it appears much more difficult and complicated than the method considered below.

10.2 Multisurface Scattering Theory

The wave functions forming the set in (10.38) can be used for functions into which a wave function for the system may be expanded in the same spirit as in Chap.5. However, unlike in the previous chapters, the wave functions $\psi_{\alpha n}^{(\pm)}$ depend explicitly on the internal coordinates in addition to the relative translational coordinates. Therefore, the space in which the functions and matrices are defined is much more extended than before, because we need a set of equations on which to base mathematical analysis of the complex trajectory method and also to start some numerical integration programs for multisurface scattering problems without resorting to an ad hoc procedure such as by *Tully* and *Preston* [10.1]. To simplify the discussion as much as possible, we specialize it to a two-potential-surface problem. It is formally straightforward to generalize it to a multipotential-surface problem. Then, the Hamiltonian takes the form

$$H = \begin{pmatrix} H_1 & V_{12} \\ V_{21} & H_2 \end{pmatrix} \tag{10.39}$$

and the corresponding wave function can be written as a column vector

$$\Psi^{(\pm)} = \begin{pmatrix} \psi_1^{(\pm)} \\ \psi_2^{(\pm)} \end{pmatrix} \quad . \tag{10.40}$$

The wave functions in (10.41) may be expressed in terms of the complete set of the WKB wave functions represented by (10.28a,38). Following the idea in Chap.5, we may formally write the wave functions in the form

$$\Psi = \alpha \Psi^{(+)} + \beta \Psi^{(-)} \quad , \tag{10.41}$$

where α and β are matrices dependent on the variables to be determined and $\Psi^{(\pm)}$ are column vectors whose elements are defined by the set of functions in (10.37). Since the vector space is spanned by the complete set of WKB wave functions for particles moving on an adiabatic surface, the dimension of space is larger than 2, although we formally write the vectors only in two components as in (10.40). Therefore, if there are n internal states for each adiabatic surface, then the expansion (10.41) consists of 4n terms.

Expansion (10.41) is similar to (5.56) and the coefficient matrices α and β may be determined similarly following the method used in Chap.5. If we further assume that there is only a single state for each adiabatic surface, then the problem becomes similar to the two-state problem discussed in Chap. 5 and the equation of motion for $X = \{\alpha_{11}, \beta_{11}, \alpha_{22}, \beta_{22}\}$ turns out to be almost the same as for the two-state problem discussed there. Therefore, to avoid repetition we briefly sketch the derivation of the equation of motion in the notation appropriate for the present problem of a two-state two-surface model. It reduces the complexity of equations, yet retains the essential features undiminished.

It is convenient to introduce a mass weighted f-dimensional gradient operator

$$D = \sum_{i=1}^{f} (2m_i)^{-1} (\partial/\partial Q_i) \quad . \tag{10.42}$$

The Schrödinger equation may then be written as

$$[D^2 + \lambda^2 (E - V)]\Psi = 0 \quad , \tag{10.43}$$

where $\lambda = 1/\hbar$.

In analogy to (5.57) we impose the condition

$$(D\alpha)\Psi^{(+)} + (D\beta)\Psi^{(-)} = 0 \quad , \tag{10.44}$$

motivated by our desire that α and β change as slowly as possible. When (10.41) is substituted into the Schrödinger equation (10.43) and condition (10.44) is used, then

$$D\alpha \cdot D\Psi^{(+)} + D\beta \cdot D\Psi^{(-)} = -[\alpha D^2 + \lambda^2 (E - V)\alpha]\Psi^{(+)} - [\beta D^2 + \lambda^2 (E - V)\beta]\Psi^{(-)} \quad . \tag{10.45}$$

Now let us assume that $\Psi^{(\pm)}$ are such that

$$D^2\Psi^{(\pm)} = -\lambda^2 \varphi^{(\pm)}\Psi^{(\pm)} \quad , \tag{10.46}$$

where $\varphi^{(\pm)}$ are diagonal matrices which may be chosen so that $\Psi_i^{(\pm)}$ become the WKB solutions determined by (10.21) or (10.30,24a), that is, the WKB solutions in the adiabatic representation. It is convenient to decompose $\varphi_i^{(\pm)}$ into two parts

$$\varphi_i^{(\pm)} = \varphi_i^{(0)} + \varphi_i^{(1,\pm)} \quad , \tag{10.47}$$

and choose $\varphi_i^{(0)}$ as one of the eigenvalues of the secular equations associated with (10.39):

$$h_i(E_i - \varphi) - \sum_{j \neq i} V_{ij}h_j = 0 \quad , \tag{10.48}$$

where $E_i = E - V_{ii}$. Clearly, the eigenvalues are the adiabatic momenta. The $\varphi_i^{(1,\pm)}$ part is then an f-dimensional analog of the Schwartzian derivatives that have shown up repeatedly before. The $\varphi^{(\pm)}$ can be explicitly found by substituting the WKB solutions into (10.46),

$$\psi_j^{(\pm)} = \Delta_j^{1/2} \exp(\pm i\lambda k_j) \quad , \quad (F_j = \pm k_j) \quad , \tag{10.49}$$

obtained with (10.21,24a), and then comparing both sides. We assume that it is done. The $\varphi_i^{(1,\pm)}$ are of the order of λ^{-2} and may be neglected in the semiclassical approximation, since the terms retained are $O(\lambda^2)$.

We now impose additional conditions equivalent to (5.58):

$$\alpha_{ik}(\varphi_k - E_i) + \sum_{j \neq i} V_{ij}\alpha_{ij} = 0 \quad ,$$

$$(k \neq i) \tag{10.50}$$

$$\beta_{ik}(\varphi_k - E_i) + \sum_{j \neq i} V_{ij}\beta_{jk} = 0 \quad ,$$

where it must be noted that $\varphi_k \equiv \varphi_k^{(\pm)}$, the kth eigenvalue of (10.48). Then (10.45) becomes

$$D\alpha \cdot D\Psi^{(+)} + D\beta \cdot D\Psi^{(-)} = \delta\alpha^{(d)}\Psi^{(+)} + \delta\alpha^{(d)}\Psi^{(-)} \quad , \tag{10.51}$$

where δ is a traceless matrix whose elements are defined by

$$\delta_{12} = \lambda^2[(\varphi_2 - E_1)\gamma_{12} + V_{12}] \quad ,$$

$$\delta_{21} = \lambda^2[(\varphi_1 - E_2)\gamma_{21} + V_{21}] \quad ,$$

$$\gamma_{12} = V_{12}^{-1}(E_1 - \varphi_1) \quad ,$$

$$\gamma_{21} = V_{21}^{-1}(E_2 - \varphi_2) \quad ,$$

and $\alpha^{(d)}$ and $\beta^{(d)}$ are the diagonal parts of the matrices α and β. Equations (10.51,44) together form a set of linear differential equations for the matrices $\alpha^{(d)}$ and $\beta^{(d)}$.

Conditions (10.50) imply that the off-diagonal elements α_{ij} and β_{ij} for $i \neq j$ can be determined in terms of the diagonal components α_{ii} and β_{ii}. Equations (5.68a,b) still apply here if the dimension of the linear space is appropriately extended and if the two-surface two-state model is replaced with a multistate model. We express the relations between the off-diagonal and diagonal components in the form

$$\alpha_{ij} = \gamma_{ij}\alpha_{jj} \;\; ; \;\; \beta_{ij} = \gamma_{ij}\beta_{jj} \quad , \quad (i \neq j) \quad . \tag{10.52}$$

Eliminating the off-diagonal elements of α and β from the linear systems (10.44,51), we obtain the following set of differential equations for the diagonal components of α and β:

$$(D\alpha^{(d)})\Psi^{(+)} + \gamma(D\alpha^{(d)})\Psi^{(+)} + (D\beta^{(d)})\Psi^{(-)} + \gamma(D\beta^{(d)})\Psi^{(-)}$$
$$= -(D\gamma)\alpha^{(d)}\Psi^{(+)} - (D\gamma)\beta^{(d)}\Psi^{(-)} \tag{10.53a}$$

$$(D\alpha^{(d)}) \cdot D\Psi^{(+)} + \gamma(D\alpha^{(d)}) \cdot D\Psi^{(+)} + (D\beta^{(d)}) \cdot D\Psi^{(-)} + \gamma(D\beta^{(d)}) \cdot D\Psi^{(-)}$$
$$= -(D\gamma)\alpha^{(d)} \cdot D\Psi^{(+)} - (D\gamma)\beta^{(d)} \cdot D\Psi^{(-)}$$
$$+ \delta\alpha^{(d)}\Psi^{(+)} + \delta\beta^{(d)}\Psi^{(-)} \quad . \tag{10.53b}$$

These equations may be cast into a more compact form similar to the equations of motion that have appeared in the previous chapters. For the purpose we define the following:

$$X = \{\alpha_{11}, \beta_{11}, \alpha_{22}, \beta_{22}\} \tag{10.54}$$

$$W_i = \begin{pmatrix} \Psi_i^{(+)} & \Psi_i^{(-)} \\ D\Psi_i^{(+)} & D\Psi_i^{(-)} \end{pmatrix} \tag{10.55}$$

$$Q = \begin{pmatrix} W_1 & \gamma_{12}W_2 \\ \gamma_{21}W_1 & W_2 \end{pmatrix} \tag{10.56}$$

$$M = - \begin{pmatrix} 0 & (D\gamma_{12})W_2 \\ (D\gamma_{21})W_1 & 0 \end{pmatrix} + \begin{pmatrix} 0 & \delta_{12}g_2 \\ \delta_{21}g_1 & 0 \end{pmatrix} \qquad (10.57)$$

$$g_j = \begin{pmatrix} 0 & 0 \\ \psi_j^{(+)} & \psi_j^{(-)} \end{pmatrix} . \qquad (10.58)$$

Then, the equation of motion for the two-surface two-state model is

$$Q \cdot DX = MX . \qquad (10.59)$$

The Wronskian determinant of the matrix W_j in (10.55) is calculated with the WKB solutions in (10.49):

$$\det |W_j| = -\Delta_j Dk_j \equiv \omega_j .$$

The inverse of the matrix W_j is

$$W_j^{-1} = \omega_j^{-1} \begin{pmatrix} D\psi_j^{(-)} & -\psi_j^{(-)} \\ -D\psi_j^{(+)} & \psi_j^{(+)} \end{pmatrix} .$$

The inverse of the matrix Q is then given in terms of the inverses of the Wronskian matrices,

$$Q^{-1} = [\omega_1\omega_2(1 - \gamma_{12}\gamma_{21})]^{-1} \begin{pmatrix} \omega_1 W_1^{-1} & -\gamma_{12}\omega_1 W_1^{-1} \\ -\gamma_{21}\omega_2 W_2^{-1} & \omega_2 W_2^{-1} \end{pmatrix} . \qquad (10.60)$$

In the classical limit of large λ we may neglect the second term in the matrix φ which contains δ_{12} and δ_{21}, in comparison with the first term that is $O(\lambda)$. In this limit we find the equation of motion

$$Dx = M_\infty X \qquad (10.61)$$

where

$$M_\infty = - \begin{pmatrix} \omega_1\gamma_{12}(D\gamma_{21})I & 0 \\ 0 & \omega_2\gamma_{21}(D\gamma_{12})I \end{pmatrix} + \begin{pmatrix} 0 & \omega_1(D\gamma_{12})P_{12} \\ \omega_2(D\gamma_{21})P_{21} & 0 \end{pmatrix} ,$$

$$\qquad (10.62)$$

$$P_{ij} = \begin{pmatrix} (\Psi_i^{(-)}, \Psi_j^{(+)}) & (\Psi_i^{(-)}, \Psi_j^{(-)}) \\ -(\Psi_i^{(+)}, \Psi_j^{(+)}) & -(\Psi_i^{(+)}, \Psi_j^{(-)}) \end{pmatrix} , \tag{10.63}$$

$$(\Psi_i^{(-)}, \Psi_j^{(+)}) = \Psi_i^{(-)} D \Psi_j^{(+)} - \Psi_j^{(+)} D \Psi_i^{(-)} , \quad \text{etc.}$$

The formal structure of these formulas is closely similar to formulas in the previous chapters. The present formulas are, however, in multidimensional space and the phase integrals and associated formulas are all defined in such space. It is instructive to consider an element of the P_{ij} matrix, say, $W(\Psi_1^{(-)}, \Psi_2^{(+)})$. When calculated explicitly with the WKB solutions defined by (10.49), the element may be written as

$$(\Psi_1^{(-)}, \Psi_2^{(+)}) = \left[\frac{1}{2} D \ln(\Delta_2/\Delta_1) + i\lambda(Dk_2 - Dk_1) \right] (\Delta_1 \Delta_2)^{1/2} \exp[i\lambda(k_2 - k_1)] .$$

Here an interesting point to observe is the phase factor which in this case consists of a difference $k_2 - k_1$ of action integrals for classical motions on two different adiabatic potential surfaces. Note also that in the equation of motion such oscillatory factors appear multiplied by a factor that depends on the parameter characterizing the coupling between the two potential surfaces. We stress that solution of the equation of motion obtained above is achieved when the classical equations of motion or the Hamilton-Jacobi equations and the amplitude equation (10.24a) are solved. Note that the latter is also a classical equation. It is also possible to continue into the complex plane to introduce a complex trajectory method, but care must be taken in doing so. Probably, it will be safe to stay on the real axis, since there is no need to go into the complex plane to get transition probabilities. The water of the complex plane is trickier than one suspects, with many shoals.

11. Scattering off an Ellipsoidal Particle: The WKB Approximation

The approximation methods described in the previous chapters are based on either Cartesian, spherical or polar coordinate systems. These coordinate systems are sometimes most natural for the problem considered, but even if they may not be the most natural for the problem, they are commonly used in molecular scattering perhaps by tradition and often for the mathematical conveniences they provide, but at the cost in other aspects. In optics and electromagnetism solutions of boundary value problems are required for systems involving certain particular geometries of the boundaries and the whole gamut of coordinate systems available in the literature is developed precisely with such situations in mind. In this chapter we consider a particular example of an ellipsoidal particle with or without a potential tail colliding with a spherical particle. If the potential tail vanishes, the problem is solvable in terms of known analytic functions of rather complicated nature, which require quite involved numerical techniques, although their computation is much studied in the literature. Nevertheless, it is useful to have an approximate method of evaluation and we shall employ the WKB method. Moreover, if there is an interaction potential, such an approximation method is inevitable. Here, we consider in detail the simplest case just to stimulate further study, but deal only briefly with the general case of scattering. There appears to be a great deal of further investigation necessary in this direction despite the fact that the use of prolate spheroidal coordinates was initiated for scattering theory from the early period of quantum mechanics [11.1-4].

Specifically, the WKB approximation is developed for the phase shifts for elastic scattering by an ellipsoid particle, which may be expressed in terms of the eccentricity of the ellipsoid, and for spheroidal harmonics [11.5-7] that appear in the expression for the cross sections in prolate spheroidal coordinates. Some aspects of inelastic scattering theory will be discussed briefly.

11.1 · Elastic Scattering off an Ellipsoid

Consider a hard ellipsoidal particle located at a point in space where the origin of the coordinate system may be located. We shall assume that the projectile is a sphere with a finite diameter. Then, by increasing the major and minor axes of the ellipsoid by the diameter of the sphere, we may regard it as a point particle. Thus, quantum mechanically the plane wave for the point projectile collides with the ellipsoid and is scattered by the latter. The problem is then to calculate the differential cross section of the wave scattered into a small surface element around a point at infinite distance from the ellipsoid. We shall assume in this section that there is no inter-action potential except that the particles are hard. We further assume that the major axis of the ellipsoid is in the direction forming an angle with the space-fixed z axis along which the plane wave is incident.

Then the Schrödinger equation is that of a free particle in the relative coordinate system:

$$-(\hbar^2/2m)\nabla^2\psi = E\psi \quad . \tag{11.1}$$

The most natural coordinates in this case are the prolate spheroidal co-ordinates and the Schrödinger equation takes the form

$$\left[\frac{\partial}{\partial\mu}\,(\mu^2 - 1)\,\frac{\partial}{\partial\mu} + \frac{\partial}{\partial\nu}\,(1 - \nu^2)\,\frac{\partial}{\partial\nu} + \frac{\mu^2 - \nu^2}{(\mu^2 - 1)(1 - \nu^2)}\,\frac{\partial^2}{\partial\phi^2}\right.$$

$$\left. + \lambda^2 p^2(\mu^2 - \nu^2)\right]\psi(\mu,\nu,\phi) = 0 \quad , \tag{11.2}$$

where

$$\mu = (r_a + r_b)/R$$

$$\nu = (r_a - r_b)/R$$

$$x = \frac{1}{2}\,R[(\mu^2 - 1)(1 - \nu^2)]^{1/2}\,\cos\phi$$

$$y = \frac{1}{2}\,R[(\mu^2 - 1)(1 - \nu^2)]^{1/2}\,\sin\phi$$

$$z = \frac{1}{2}\,R\mu\nu \quad ,$$

and the volume element is

$$dv = (R^3/8)(\mu^2 - \nu^2)d\mu\;d\nu\;d\phi$$

with the ranges of the variables defined by

$$1 \le \mu < \infty, \quad -1 \le \nu \le 1 \;, \quad 0 \le \phi \le 2\pi \quad .$$

Other notations in (11.2) are

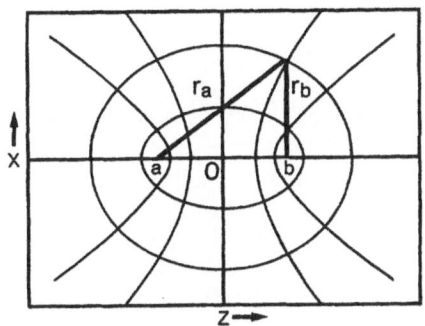

Fig.11.1.
Prolate spheroidal coordinates

$$\lambda = (2m)^{1/2}/\hbar$$

$$p = RE^{1/2} \quad ,$$

where R is the distance between the foci, and r_a and r_b are indicated in
Fig.11.1. We now assume that the ellipsoid has a size defined by $\mu_0 > 1$
(Note that the ellipsoid is obtained by rotating the ellipse μ_0 around the
z axis).

If the wave function is written in the form

$$\Psi(\mu,\nu,\phi) = J(\mu)S(\nu)\Phi(\nu) \quad ,$$

(11.2) is separable into three differential equations,

$$\left[\frac{d}{d\mu} (\mu^2 - 1) \frac{d}{d\mu} - \frac{m^2}{\mu^2 - 1} + \lambda^2 p^2 \mu^2 - \tau_{m\ell}\right] J(\mu) = 0 \tag{11.3}$$

$$\left[\frac{d}{d\nu} (1 - \nu^2) \frac{d}{d\nu} - \frac{m^2}{1 - \nu^2} - \lambda^2 p^2 \nu^2 + \tau_{m\ell}\right] S(\nu) = 0 \tag{11.4}$$

$$\frac{d^2}{d\phi^2} \Phi(\phi) = -m^2 \Phi(\phi) \quad , \tag{11.5}$$

where m and $\tau_{m\ell}$ are separation constants. The subscripts m and ℓ of ι denote
that the separation constant is quantized and dependent on m and ℓ. It will
turn out to be the eigenvalue of spheroidal wave function $S_{m\ell}$. Equation (11.5)
is easily solved:

$$\Phi_m(\phi) = (2\pi)^{-1/2} \exp(im\phi) \quad , \quad (m = 0,\pm1,\pm2,\dots) \quad . \tag{11.6}$$

It is known that the solution for (11.4) can be given in terms of associated
Legendre polynomials [11.5-7]

$$S_{m\ell}(\lambda p,\nu) = \sum_n{}' d_n(\lambda p|m\ell) P_{n+m}^m(\nu) \quad , \tag{11.7}$$

where d_n are coefficients obeying a three-term recursion formula [11.5] and
the prime on the summation sign indicates that the values of n are even if
(ℓ - m) is even, and odd if (ℓ - m) is odd. The index ℓ is an integer such

that

$\ell = m, m + 1, m + 2, \ldots$.

The separation constant $\tau_{m\ell}$ is given in terms of ℓ and m as well as p. The leading values of $\tau_{m\ell}$ are as follows:

$$\tau_{00} \simeq (\lambda p)^2/3 - 2(\lambda p)^4/135$$

$$\tau_{01} \simeq 2 + 3(\lambda p)^2/5 - 6(\lambda p)^4/875 \quad .$$

The eigenfunctions $S_{m\ell}$ are orthogonal and the normalization constant is denoted by $\Lambda_{m\ell}(\lambda p)$:

$$\Lambda_{m\ell}(\lambda p) = \int_{-1}^{1} d\nu |S_{m\ell}(\lambda p,\nu)|^2$$

$$= \sum_{n}{}' [d_n(\lambda p|m\ell)]^2 [2(n + 2m)!/n!(2n + 2m + 1)] \quad . \tag{11.8}$$

It is useful to note that

$$\lim_{\nu \to 1} S_{m\ell}(\lambda p,\nu) = \delta_{m0} \quad . \tag{11.9}$$

The regular solution of (11.3) takes the following form:

$$je_{m\ell}(\lambda p,\mu) = \frac{(\ell - m)!}{(\ell + m)!} \left(\frac{\mu^2 - 1}{\mu^2}\right)^{m/2} \sum_{n}{}' i^{m+n-\ell} d_n(\lambda p|m\ell)$$

$$\times \frac{(n + 2m)!}{n!} j_{n+m}(\lambda p\mu) \quad , \tag{11.10}$$

where $j_a(z)$ are spherical Bessel functions.

The irregular solution of (11.3) may be given in terms of spherical Neumann functions $n_a(z)$:

$$ne_{m\ell}(\lambda p,\mu) = \frac{(\ell - m)!}{(\ell + m)!} \left(\frac{\mu^2 - 1}{\mu^2}\right)^{m/2} \sum_{n}{}' i^{m+n-\ell} d_n(\lambda p|m\ell)$$

$$\times \frac{(n + 2m)!}{n!} n_{n+m}(\lambda p\mu) \quad . \tag{11.11}$$

These functions are analogs of spherical Bessel and Neumann functions to which they tend as $\lambda p \to 0$. As $\lambda p \to \infty$, they have the following asymptotic formulas:

$$je_{m\ell}(\lambda p,\mu) \to (\lambda p\mu)^{-1} \cos[\lambda p\mu - \tfrac{1}{2}\pi(\ell + 1)] \quad ,$$

$$ne_{m\ell}(\lambda p,\mu) \to (\lambda p\mu)^{-1} \sin[\lambda p\mu - \tfrac{1}{2}\pi(\ell + 1)] \quad . \tag{11.12a}$$

These asymptotic formulas were calculated by *Morse* and *Feshbach* [11.5].
They may be improved somewhat by replacing $(\ell + 1)\pi/2$ with $(\tau_{m\ell}^{\frac{1}{2}}\pi/2 + \pi/4)$:

$$je_{m\ell}(\lambda p,\mu) \sim (\lambda p\mu)^{-1} \cos(\lambda p\mu - \tau_{m\ell}^{\frac{1}{2}}\pi/2 - \pi/4)$$

$$ne_{m\ell}(\lambda p,\mu) \sim (\lambda p\mu)^{-1} \sin(\lambda p\mu - \tau_{m\ell}^{\frac{1}{2}}\pi/2 - \pi/4) \quad . \tag{11.12b}$$

This replacement is intuitively reasonable since $\tau_{m\ell}(\mu^2 - 1)^{-1}$ in the radial
equation plays the role of the centrifugal potential, played by $\ell(\ell + 1)r^{-2}$
in the differential equation for the spherical Bessel functions having the
same asymptotic formulas as in (11.12a). This may be justified mathematically
by using the WKB method, as will be shown later.

It is useful to introduce the analog of the spherical Hankel function
defined by

$$he_{m\ell}(\lambda p,\mu) = je_{m\ell}(\lambda p,\mu) + ine_{m\ell}(\lambda p,\mu) \tag{11.13}$$

which behaves asymptotically like

$$he_{m\ell}(\lambda p,\mu) \to i^{-\ell-1}(\lambda p\mu)^{-1} \exp(i\lambda p\mu) \quad , \tag{11.14}$$

if the asymptotic formulas in (11.12a) are used.

The radial part of the wave function is therefore given by a linear combination of two independent solutions. We may take it in the form

$$\psi_{m\ell}(\mu) = Aje_{m\ell}(\lambda p,\mu) + Bhe_{m\ell}(\lambda p,\mu) \quad . \tag{11.15}$$

Since this wave function must vanish at the boundary $\mu = \mu_0 > 1$ according
to the hard ellipsoid assumption made, then

$$Aje_{m\ell}(\lambda p,\mu_0) + Bhe_{m\ell}(\lambda p,\mu_0) = 0 \quad . \tag{11.16}$$

By introducing the amplitude and phase functions by the relations

$$je_{m\ell}(\lambda p,\mu_0) = A_{m\ell}(\lambda p,\mu_0) \sin\eta_{m\ell}(\lambda p,\mu_0)$$

$$ne_{m\ell}(\lambda p,\mu_0) = A_{m\ell}(\lambda p,\mu_0) \cos\eta_{m\ell}(\lambda p,\mu_0) \quad ,$$

we find from (11.16)

$$B = -iA \exp(i\eta_{m\ell}) \sin\eta_{m\ell} \quad . \tag{11.17}$$

Therefore, substitution into (11.15) yields

$$\psi_{m\ell}(\mu) = A[je_{m\ell}(\lambda p,\mu) - i \exp(i\eta_{m\ell}) \sin\eta_{m\ell} \, he_{m\ell}(\lambda p,\mu)] \quad . \tag{11.18}$$

The incident plane wave propagating in the direction of (ν_0,ϕ_0) can be
expanded in terms of spheroidal wave functions [11.5],

$$\exp(i\mathbf{k}\cdot\mathbf{r}) = 2 \sum_{m\ell} i^{\ell} T_m [\Lambda_{m\ell}(\lambda p)]^{-1} S_{m\ell}(\lambda p,\nu) S_{m\ell}(\lambda p,\nu_0)$$

$$\times \cos[m(\phi - \phi_0)] je_{m\ell}(\lambda p,\mu) \quad , \tag{11.19}$$

where $T_m = 1$ for $m = 0$ and 2 for $m \neq 0$. The above expansion is the prolate spheroidal coordinate counterpart of the plane-wave expansion in the spherical coordinates. In the special case of $m = 0$ the expansion looks very much like the latter.

The wave function may be expanded in terms of $\psi_{m\ell}(\mu)$,

$$\Psi(\mu,\nu,\phi) = \sum_{m\ell} A_{m\ell} S_{m\ell}(\lambda p,\nu) \cos[m(\phi - \omega_{m\ell})]$$

$$\times [je_{m\ell}(\lambda p,\mu) + i \sin n_{m\ell} \exp(in_{m\ell}) he_{m\ell}(\lambda p,\mu)] \quad , \tag{11.20}$$

where $\omega_{m\ell}$ are undetermined phase factors which we may choose along with $A_{m\ell}$ by comparing (11.20) with (11.19) according to the standard procedure in scattering theory. We find that $\omega_{m\ell}$ and $A_{m\ell}$ may be respectively taken as

$$\omega_{m\ell} = \phi_0$$

and

$$A_{m\ell} = 2i T_m [\Lambda_{m\ell}(\lambda p)]^{-1} S_{m\ell}(\lambda p,\phi_0) \quad .$$

The second part on the right-hand side of (11.20) represents the outgoing scattered partial waves which behave asymptotically as

$$\Psi_{scattered} = \sum_{m\ell} i C_{m\ell} S_{m\ell}(\lambda p,\nu) \cos[m(\phi - \phi_0)]$$

$$\times \sin n_{m\ell} \exp(in_{m\ell}) he_{m\ell}(\lambda p,\mu)$$

$$\sim F(\lambda p|\nu,\phi;\nu_0,\phi_0)(2/R\mu) \exp(i\lambda p\mu) \quad , \tag{11.21}$$

where the scattering amplitude is given by

$$F(\lambda p|\nu,\phi;\nu_0,\phi_0) = (R/4i\lambda p) \sum_{m\ell} 2 T_m [\Lambda_{m\ell}(\lambda p)]^{-1} S_{m\ell}(\lambda p,\nu)$$

$$\times S_{m\ell}(\lambda p,\nu_0) \cos[m(\phi - \phi_0)][\exp(2in_{m\ell}) - 1] \quad . \tag{11.22}$$

Note that by definition $4i\lambda p = 2ikR$. The scattering amplitude (11.22) is the analog of the scattering amplitude in the spherical coordinate system.

If the direction of the plane-wave propagation is fixed at $(\nu_0,\phi_0) = (0,0)$, then since the angular spheroidal wave function $S_{m\ell}$ has the property

$$S_{m\ell}(\lambda p,0) = \delta_{m0} \quad ,$$

the following expression is obtained from (11.22) for the scattering amplitude:

$$F(\lambda p|\nu,\phi;0,0) = (ik)^{-1} \sum_{\ell=0}^{\infty} 2[\Lambda_{0\ell}(\lambda p)]^{-1} S_{0\ell}(\lambda p,\nu)[\exp(2i\eta_{0\ell}) - 1] \quad ,$$

(11.23)

which is independent of ϕ. This is so because the ellipsoid is aligned along the z axis and it is symmetric about its major axis.

The scattering cross section $\sigma(\nu,\phi|\nu_0,\phi_0)$ may be defined by the relation

$$\sigma(\nu,\phi|\nu_0,\phi_0)d\nu\ d\phi = |F(\lambda p|\nu,\phi;\nu_0,\phi_0)(2/R\mu)|^2(R^2/4)(\mu^2 - \nu^2)d\nu\ d\phi$$

$$= |F(\lambda p|\nu,\phi;\nu_0,\phi_0)|^2[1 + O(\mu^{-2})]d\nu\ d\phi$$

(11.24)

at large μ. The observable cross section $\sigma(\nu,\phi|\lambda p)$ is therefore obtained if the cross sections at various orientations (ν_0,ϕ_0) above are integrated over their ranges:

$$\sigma(\nu,\phi|\lambda p) = \int_0^{2\pi} d\phi_0 \int_{-1}^1 d\nu_0\ \sigma(\nu,\phi|\nu_0,\phi_0)$$

$$= \int_0^{2\pi} d\phi_0 \int_{-1}^1 d\nu_0|F(\lambda p|\nu,\phi;\nu_0,\phi_0)|^2 \quad .$$

(11.25)

This is the elastic scattering cross section by the ellipsoid. The formula is quite reminiscent of the scattering cross section for a spherical particle. In place of Legendre polynomials spheroidal harmonics occur and the phase shifts depend on the quantum number m in addition to ℓ. For the hard ellipsoid system considered here the phase shifts are in principle determined by

$$\eta_{m\ell}(\lambda p,\mu_0) = \tan^{-1}[je_{m\ell}(\lambda p,\mu_0)/ne_{m\ell}(\lambda p,\mu_0)] \quad .$$

(11.26)

If m and ℓ are both small, this form for $\eta_{m\ell}$ is straightforward to compute, but as the quantum numbers increase, the computation becomes increasingly difficult. The phase shifts may be computed by an asymptotic expansion method.

11.2 The WKB Approximation

The wave functions $je_{m\ell}$ and $ne_{m\ell}$ and phase shifts $\eta_{m\ell}$ can be computed by solving a three-term recursion formula for the coefficients $d_n(\lambda p|m\ell)$. The recursion formula is usually solved by casting it in a continued fraction and truncating the latter [11.6]. The scheme works fine for small ℓ, but becomes impractical as ℓ gets large. Unfortunately, a large number of ℓ is required to calculate the cross section to a desired accuracy. Here we shall consider the WKB approximations for the wave functions.

We first consider the radial part of the wave function. If we set

$$J_{m\ell}(\mu) = (\mu^2 - 1)^{-1/2} R_{m\ell}(\mu) \quad , \tag{11.27}$$

then the new function $R_{m\ell}$ satisfies the differential equation

$$[d^2/d\mu^2 + \lambda^2 Q_{m\ell}(\mu)] u_{m\ell}(\mu) = 0 \tag{11.28}$$

$$\lambda^2 Q_{m\ell}(\mu) = \lambda^2 p^2 - \lambda^2 V_{eff}(\mu) \quad , \tag{11.29}$$

where the effective potential $V_{eff}(\mu)$ is defined

$$\lambda^2 V_{eff}(\mu) = \frac{m^2 - 1}{(\mu^2 - 1)^2} + \frac{\tau_{m\ell}}{\mu^2 - 1} - \frac{\lambda^2 p^2}{\mu^2 - 1} \quad . \tag{11.30}$$

Henceforth we shall replace $m^2 - 1$ in the effective potential with $(m + d)^2$, $d \geq 0$, in an analogy to the Langer modification [2.18b] which replaces $\ell(\ell + 1)$ with $(\ell + 1/2)^2$ in the radial Schrödinger equation (Chap.4 and Sect. 11.4). The significance of the constant d is discussed in Sect.11.4 where the uniform WKB method is developed for $S_{m\ell}$. This replacement is consistent with the procedure used in the WKB approximation for spheroidal harmonics in Sect.11.4. In effect, it eliminates the nonbijectivity of transformations which arises when $m = 0$, since then the potential changes from a repulsive to an attractive potential if the modification is not made. Although the bijectivity argument does not require that d be nonzero in the case of the radial equation we are considering here, we take the same value of d as in the case of $S_{m\ell}$ for the sake of consistency. The choice of the value of d will be discussed in detail in Sect.11.5.

The WKB solution for (11.28) is then

$$R_{m\ell}(\mu) = A Q_{m\ell}^{-1/4}(\mu) \exp\left[i\lambda \int_{\mu_t}^{\mu} dt \; Q_{m\ell}^{1/2}(t) \right]$$

$$+ B Q_{m\ell}^{-1/4}(\mu) \exp\left[-i\lambda \int_{\mu_t}^{\mu} dt \; Q_{m\ell}^{1/2}(t) \right] \quad , \tag{11.31}$$

where μ_t is the turning point (Fig.11.2). When the boundary condition is imposed at $\mu = \mu_t$ and an appropriate connection formula is used (Chap.2), we obtain the solution in the form

$$R_{m\ell}(\mu) = A Q_{m\ell}^{-1/4}(\mu) \cos\left[\lambda \int_{\mu_t}^{\mu} dt \; Q_{m\ell}^{1/2}(t) - \pi/4 \right] \tag{11.32}$$

which behaves asymptotically like

.

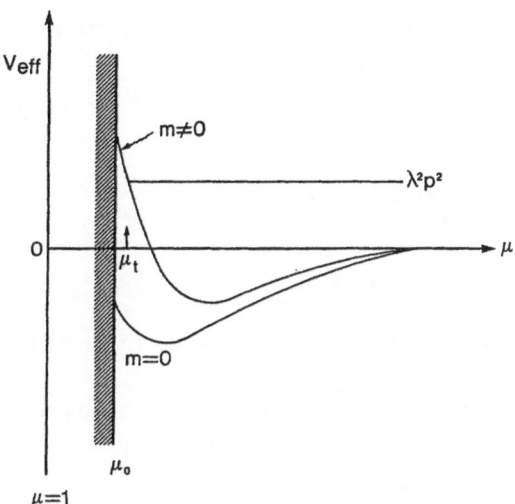

Fig.11.2. Effective potentials for the radial spheroidal wave equation

$$R_{m\ell}(\mu) \sim A(\lambda p)^{-1} \cos(\lambda p\mu - \pi\tau_{m\ell}^{\frac{1}{2}}/2 - \pi/4 + n_{m\ell}) \quad . \tag{11.33}$$

Here the phase shift $n_{m\ell}$ is given by

$$n_{m\ell} = \pi\tau_{m\ell}^{\frac{1}{2}}/2 - \lambda p\mu_t + \lambda \int_{\mu_t}^{\infty} dt[Q_{m\ell}^{\frac{1}{2}}(t) - p] \quad , \tag{11.34}$$

and it may be identified with $n_{m\ell}$ in the previous section. Therefore, the phase shifts appearing in (11.23) are now given by

$$n_{0\ell} = \pi\tau_{0\ell}^{\frac{1}{2}}/2 - \lambda p\mu_t + \lambda \int_{\mu_t}^{\infty} [Q_{0\ell}^{\frac{1}{2}}(t) - p] \quad . \tag{11.35}$$

These look similar to the WKB phase-shift formula in the spherical coordinate representation.

The phase shifts can be expressed in terms of elliptic integrals. We shall consider $n_{0\ell}$ for $d = 0$ especially since this case turns out to be quite simple and illuminating. The result for $n_{m\ell}$ ($m \neq 0$ and $d \neq 0$) is not as simple, consisting of a number of elliptic integrals. There are two distinct cases to consider: $\tau_{0\ell} > \lambda^2 p^2$ and $\tau_{0\ell} < \lambda^2 p^2$.

I) *The case of* $\tau_{0\ell} > \lambda^2 p^2$. In this case the effective potential is repulsive everywhere in the range of μ. The classical turning point is then

$$\mu_t = \tau_{0\ell}^{\frac{1}{2}}/\lambda p \equiv \gamma_\ell^{-1} \quad .$$

The integral in (11.35) can be expressed in terms of an elliptic integral of the second kind and the phase shift takes the form

$$\eta_{0\ell} = (\lambda p/\gamma_\ell)[\pi/2 - E(\pi/2|\gamma_\ell)] \quad , \tag{11.36}$$

where

$$E(\pi/2|\gamma_\ell) = \int_0^1 dx(1 - \gamma_\ell^2 x^2)^{1/2}(1 - x^2)^{-1/2}$$

$$= \frac{\pi}{2}\left(1 - \tfrac{1}{4}\gamma_\ell^2 - \frac{3}{64}\gamma_\ell^4 - \ldots\right)(\gamma_\ell < 1) \quad .$$

Since γ_ℓ vanishes as ℓ tends to infinity, the phase shift tends to zero in the limit.

It is useful to cast the phase-shift formula in terms of the eccentricity of the ellipsoid, since such a mode of expression will clearly exhibit the difference between the phase shifts of the hard sphere and ellipsoid. The eccentricity of an ellipsoid of size μ_t is defined by the parameter

$$\varepsilon_\ell = 1/\mu_t \quad ,$$

or by the ratio of the semiminor axis b_ℓ to the semimajor axis a_ℓ

$$\varepsilon_\ell = [1 - (b_\ell/a_\ell)^2]^{1/2} \quad .$$

Then, since $\lambda p \mu_t = ka_\ell$ and hence

$$ka_\ell = kb_\ell(1 - \varepsilon_\ell^2)^{-1/2} \quad ,$$

we see that the phase shifts can be expressed explicitly as a function of the eccentricity. In fact, the parameter γ_ℓ is simply the eccentricity. The phase shifts may be written as

$$\eta_{0\ell} = kb_\ell(1 - \varepsilon_\ell^2)^{-1/2}[\pi/2 - E(\pi/2|\varepsilon_\ell)] \quad . \tag{11.37}$$

If we assume that b_ℓ is the radius of a sphere that the molecule effectively assumed if it were taken to be a sphere, then we may define the phase shift η_ℓ for the hard sphere by the well-known expression

$$\eta_\ell = \pi(\ell + 1/2)/2 - kb_\ell \quad . \tag{11.38}$$

We can determine the nonsphericity correction due to the ellipsoidal nature of the molecule from the difference between (11.37,38).

II) *The case of* $\tau_{0\ell} < \lambda^2 p^2$. Since the effective potential is purely attractive in this case, the classical turning point is μ_0 and independent of ℓ. The phase shifts in the present case may be expressed in terms of elliptic integrals as follows:

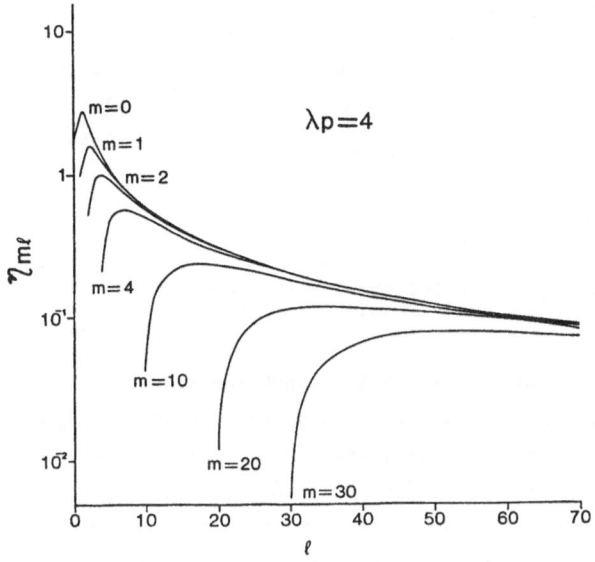

Fig.11.3. Phase shifts
versus ℓ at various values
of m

$$\eta_{0\ell} = \pi\tau_{0\ell}^{1/2}/2 - \lambda p\mu_0 + \lambda p\bar{\Delta}_{0\ell} \quad , \tag{11.39}$$

where with $c_\ell = \tau_{0\ell}^{1/2}/\lambda p < 1$

$$\bar{\Delta}_{0\ell} = \left\{ F(\pi/2|c_\ell) - F(\varphi_0|c_\ell) - E(\pi/2|c_\ell) + E(\varphi_0|c_\ell) - c_\ell^2 F(\sin^{-1}\varepsilon_0|c_\ell) \right.$$

$$\left. - \varepsilon_0 \left\{ \left[(1 - \varepsilon_0^2)/(1 - c_\ell^2\varepsilon_0^2) \right]^{1/2} - 1 \right\} \right. \tag{11.40}$$

$$\varepsilon_0 = 1/\mu_0$$

$$\varphi_0 = \sin^{-1}\left[(1 - \varepsilon_0^2)/(1 - c_\ell^2\varepsilon_0^2) \right]^{1/2} \quad .$$

Here $F(\varphi|c_\ell)$ denotes an elliptic integral of the first kind [2.2,14]. The
phase-shift formulas given above can be easily computed as a function of ℓ.

Figure 11.3 shows its ℓ and m dependence at a given value of λp (d $\neq$ 0).
Unlike the phase shifts for a hard sphere which are negative everywhere,
the phase shifts for an ellipsoid behave like those for particles interacting through a potential with an attractive branch, say, the Lennard-Jones
(12,6) potential. This effect is not intuitively obvious, but very interest-
ing. Since $\ell \geq |m|$, the phase shifts start from $\ell \geq |m|$, but they all vanish
eventually to zero, although the rate of decrease is rather slow due to the
long-range nature of the effective potentials.

11.3 Elastic Scattering by a Potential

The results in Sect.11.2 can be generalized to include elastic scattering
through a potential with a finite range. To keep the formalism simple, we
assume a special type of potential. Specifically, the potential is assumed
to take the form

$$V(r_a, r_b) = V(r_a + r_b)/r_a r_b \quad , \tag{11.41a}$$

which in the prolate spheroidal coordinates assumes the form

$$V = V(\mu)/[R^2(\mu^2 - \nu^2)] \quad . \tag{11.41b}$$

We of course assume that $\lim_{\mu \to 1} V(\mu) = \infty$. We shall absorb the factor R^2 into
$V(\mu)$ in the subsequent discussion. The potential assumed is clearly anisotropic from the viewpoint of spherical coordinates. Nevertheless, the prolate
spheroidal coordinates behave like the spherical coordinates, and in particular the μ coordinate approaches the radial coordinate in the spherical coordinate system, as μ increases. Therefore, the potential assumed above asymptotically becomes spherically symmetric at large μ. A useful model for $V(\mu)$
may be taken in the form

$$V(\mu) = V_0[(\sigma/\mu)^n - (\sigma/\mu)^{n'}] \quad , \tag{11.41c}$$

where V_0, σ, n, and n' are parameters. For example, if we choose n = 12 and
n' = 6, the model formally takes the same form as the Lennard-Jones (12,6)
potential to which it reduces as $\eta \to \infty$. Note that this potential is intrinsically anisotropic.

The Schrödinger equation is then

$$[-(\hbar^2/2m)\nabla^2 + V]\Psi = E\Psi \quad ,$$

which in prolate-spheroidal coordinates separates into three independent
differential equations similar to (11.3-5). In fact, the angular parts are
exactly the same as (11.4,5) and the radial equation is slightly modified
due to the presence of the potential:

$$\left\{ \frac{d}{d\mu} (\mu^2 - 1) \frac{d}{d\mu} - \frac{m^2}{\mu^2 - 1} + \lambda^2[p^2\mu^2 - V(\mu)] - \tau_{m\ell} \right\} J_{m\ell}(\mu) = 0 \quad , \tag{11.42}$$

where subscripts m and ℓ of τ and J(μ) indicate that they are associated
with the angular parts of the wave function, $S_{m\ell}$ and Φ_m. The total wave function is therefore given in the form

$$\Psi(\mu, \nu, \phi) = J_{m\ell}(\mu) \, S_{m\ell}(\nu) \, \Phi_m(\phi) \quad .$$

Equation (11.42) may be solved by means of the WKB method. Since it is completely parallel to what is discussed in Sect.11.2, we just present the result:

$$J_{m\ell}(\mu) = (\mu^2 - 1)^{-1/2} R_{m\ell}(\mu) \quad ,$$

where

$$R_{m\ell}(\mu) = A \, q_{m\ell}^{-\frac{1}{4}}(\mu) \, \cos\left[\lambda \int_{\mu_m}^{\mu} dt \, q_{m\ell}^{\frac{1}{2}}(t) - \pi/4\right] \quad , \tag{11.43}$$

$$\lambda^2 q_{m\ell}^2(\mu) = \lambda^2 p^2 - \lambda^2 V_{eff}(\mu) - \lambda^2(\mu^2 - 1)^{-1} V(\mu) \quad , \tag{11.44}$$

where $V_{eff}(\mu)$ is now defined as follows:

$$\lambda^2 V_{eff}(\mu) = \frac{m^2}{(\mu^2 - 1)^2} + \frac{\tau_{m\ell} - \lambda^2 p^2}{\mu^2 - 1} + \frac{\lambda^2 V(\mu)}{\mu^2 - 1} \quad . \tag{11.30a}$$

The turning point μ_t is defined by

$$q_{m\ell}(\mu_t) = 0 \quad .$$

Following the same procedure as for (11.34) for the hard-ellipsoid phase shifts, we obtain from (11.44) the phase shifts

$$\eta_{m\ell} = \frac{\pi}{2} \, \tau_{m\ell}^{\frac{1}{2}} - \lambda p \mu_t + \lambda \int_{\mu_t}^{\infty} dt [q_{m\ell}^{\frac{1}{2}}(t) - p] \quad . \tag{11.45}$$

This phase shift equation has the same formal structure as that for a hard ellipsoid. This formula goes into (11.22 or 23) for the scattering amplitude.

11.4 The WKB Approximation for Spheroidal Harmonics

In contrast to their spherical coordinate counterparts, the phase shifts in the prolate spheroidal coordinates explicitly depend on the magnetic quantum number in addition to the orbital angular momentum quantum number. Moreover, their m and ℓ dependences as presented are implicit, since they appear through the eigenvalues $\tau_{m\ell}$ which are not precisely known until the spheroidal wave equation (11.4) is solved. As mentioned earlier, the eigenvalue problem requires the solution of a three-term recursion relation, which is generally not solvable in closed form. Stratton et al. [11.6] tabulated the spheroidal wave functions for small m and ℓ, but the spheroidal wave functions for large m and ℓ are required for calculating cross sections. Therefore, it is necessary to consider an approximation method to treat the eigenvalue problem for large m and ℓ.

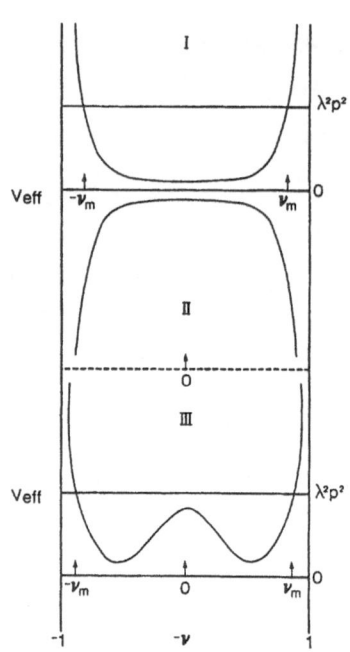

Fig.11.4. Effective potentials for the
angular spheroidal wave equation

Consequently, if we set

$$S_{m\ell}(\lambda p, \nu) = (1 - \nu^2)^{-1/2} T_{m\ell}(\lambda p, \nu) \quad , \tag{11.46}$$

then $T_{m\ell}$ obeys the differential equation

$$[d^2/d\nu^2 + \lambda^2 g_{m\ell}(\nu)] T_{m\ell}(\lambda p, \nu) = 0 \quad , \tag{11.47}$$

where

$$\lambda^2 g_{m\ell}(\nu) = \lambda^2 p^2 + \frac{\tau_{m\ell} - \lambda^2 p^2}{1 - \nu^2} - \frac{m^2 - 1}{(1 - \nu^2)^2} \quad , \tag{11.48}$$

for which we may define an effective potential $V_{m\ell}(\nu)$

$$\lambda^2 V_{m\ell}(\nu) = \frac{m^2 - 1}{(1 - \nu^2)^2} + \frac{\lambda^2 p^2 - \tau_{m\ell}}{1 - \nu^2} \tag{11.48a}$$

whose behavior is schematically described in Fig.11.4. Note that $m^2 - 1$ will
eventually be replaced by $(m + d)^2$.

In the case of $\lambda^2 p^2 - \tau_{m\ell} > 0$ it looks like a flattened parabola and in
the case of $\lambda^2 p^2 - \tau_{m\ell} < 0$ it is a double-well potential. In any case,
(11.47) requires solution of a two-turning-point problem. (In the case of
bound states four turning points can arise, since $\lambda^2 p^2$ can become lower than
the maximum of the effective potential at $\nu = 0$.)

Equation (11.47) may be solved by means of the WKB method. But in view of the fact that this solution is needed for calculating the differential cross sections at all angles, i.e., at all values of $\nu = \cos\theta$, but the WKB solution would break down at the turning points, it is preferable to have a uniform WKB solution. To implement this idea, we first consider the differential equation for associated Legendre polynomials

$$\left[\frac{d}{d\nu}(1 - \nu^2)\frac{d}{d\nu} + \ell(\ell + 1) - \frac{m^2}{1 - \nu^2}\right]P_\ell^m(\nu) = 0 \quad . \tag{11.49}$$

By setting

$$P_\ell^m(\nu) = (1 - \nu^2)^{-1/2}W_{\ell m}(\nu) \quad ,$$

we obtain the differential equation for $W_{\ell m}$

$$\left[\frac{d^2}{d\nu^2} + \frac{\ell(\ell + 1)}{1 - \nu^2} - \frac{m^2 - 1}{(1 - \nu^2)^2}\right]W_{\ell m}(\nu) = 0 \quad . \tag{11.50}$$

This is to be compared with (11.47) for $T_{m\ell}$. Notice that $T_{m\ell}$ approaches $W_{\ell m}$ as $\lambda p \to 0$, since according to *Stratton* et al. [11.6]

$$\lim_{p \to 0} \tau_{m\ell}(p) = \lim_{p \to 0} \left\{\ell(\ell + 1) + \lambda^2 p^2 \left[\frac{2\ell(\ell + 1) - 2m^2 - 1}{(2\ell - 1)(2\ell + 3)} + t_{m\ell}(\lambda p)\right]\right\}$$

$$= \ell(\ell + 1) \quad ,$$

where $t_{m\ell}$, which is a function of m, ℓ, and λp, is generally small.

We follow the idea for the uniform WKB solutions used in the previous chapters and rewrite (11.47) in the form

$$[d^2/d\nu^2 + \lambda^2 f(t)t'^2 + \tfrac{1}{2}\{t,\nu\}]T_{m\ell}$$

$$= [\lambda^2 f(t)t'^2 - \lambda^2 g_{m\ell} + \tfrac{1}{2}\{t,\nu\}]T_{m\ell}, \quad \left(t' = \frac{dt}{d\nu}\right) ,\tag{11.51}$$

where t is a new variable. Then if we choose $f(t)$ such that

$$f(t) = \frac{c^2}{1 - t^2} - \frac{\gamma^2}{(1 - t^2)^2} \tag{11.52}$$

and moreover if

$$f(t)\left(\frac{dt}{d\nu}\right)^2 = g_{m\ell}(\nu) \quad , \tag{11.53}$$

the new variable t is defined in terms of the old variable ν. Furthermore, if we set

$$T_m = \left(\frac{dt}{d\nu}\right)^{-1/2} Y(t) \tag{11.54}$$

and ignore the Schwartzian derivative $\{t,\nu\}/2$ on the right-hand side of (11.51), then the differential equation for $Y(t)$ takes the form

$$\left[\frac{d^2}{dt^2} + \lambda^2\left(\frac{c^2}{1-t^2} - \frac{\gamma^2}{(1-t^2)^2}\right)\right] Y(t) = 0 \quad . \tag{11.55}$$

The solution of (11.55) must be finite at the boundaries at $t = \pm 1$. This boundary condition is satisfied if the constants C and γ are such that

$$\lambda^2 c^2 = \ell(\ell + 1)$$

$$\lambda^2 \gamma^2 = m^2 - 1 \quad ,$$

where ℓ and m are integers subject to the condition

$$-\ell \le m \le \ell \quad .$$

In fact, comparing (11.55) with (11.50) suggests that

$$Y(t) = (1 - t^2)^{1/2} P_\ell^m(t) \tag{11.56}$$

with the variable now being the new variable t. Therefore, we finally obtain

$$T_{m\ell}(\nu) = (1 - t^2)^{1/2} (t')^{-1/2} P_\ell^m(t) \quad , \tag{11.57a}$$

so that the uniform approximation of $S_{m\ell}$ is

$$S_{m\ell}(\lambda p,\nu) = N_{m\ell}\left(\frac{1-t^2}{1-\nu^2}\right)^{1/2} (t')^{-1/2} P_\ell^m(t)$$

$$\equiv Y_{\ell m}(\lambda p,\nu) \quad . \tag{11.57b}$$

The prime on t denotes the derivative with respect to ν in the equations above and $N_{m\ell}$ is the normalization constant of the approximate uniform semiclassical spheroidal wave function $Y_{\ell m}(\lambda p,\nu)$:

$$N_{m\ell}^{-2} = \int_{-1}^{1} d\nu \left(\frac{1-t^2}{1-\nu^2}\right)^{1/2} (t')^{-1} [P_\ell^m(t)]^2 \quad . \tag{11.57c}$$

It is easy to see from the transformation (11.53) that the semiclassical approximation for $S_{m\ell}$ as given in (11.57b) does not have singularities at the turning points and thus is uniform: it is the desired result.

There now remain the questions of bijectivity of the transformation (11.53) and the determination of $\tau_{m\ell}$ in terms of α, ℓ, m, and λp. Let us consider the bijectivity first. Since the function $f(t)$ is symmetric with respect to the transformation $t \to -t$, the two zeros of $f(t)$ are symmetric around $t = 0$. We denote them by t_0 and $-t_0$. Consider the case when $\lambda^2 p^2 - \tau_{m\ell} \geq 0$. Then there are two real roots of $g_{m\ell}$, say, $-\nu_0$ and ν_0. Therefore there appears to be a one-to-one correspondence between the turning points $(-t_0, t_0)$ and $(-\nu_0, \nu_0)$. That is, the transformation is bijective. Then we integrate (11.53) to obtain

$$\int_{-t_0}^{t} dx \ f^{1/2}(x) = \int_{-\nu_0}^{\nu} dy \ g_{m\ell}^{1/2}(y) \quad . \tag{11.58}$$

We check if the transformation is indeed bijective.

For the purpose we write t' in the form

$$t'(\nu) = \frac{\lambda p(1 - t^2)}{[\ell(\ell + 1)]^{\frac{1}{2}}(1 - \nu^2)} \ \frac{(1 - \nu^2)^2 - \alpha^2(1 - \nu^2) - \beta^2}{(1 - t^2) - \gamma^2} \quad , \tag{11.59}$$

where

$$\alpha^2 = (\lambda^2 p^2 - \tau_{m\ell})/\lambda^2 p^2$$

$$\beta^2 = (m^2 - 1)/\lambda^2 p^2$$

$$\gamma^2 = (m^2 - 1)/\ell(\ell + 1) \quad . \tag{11.60}$$

Equation (11.59) clearly shows both the turning-point and singular-point structure of $g_{m\ell}(\nu)$ and $f(t)$. The turning points $\pm\nu_0$ and $\pm t_0$ are roots of the equations

$$(1 - \nu^2)^2 - \alpha^2(1 - \nu^2) - \beta^2 = 0$$

$$(1 - t^2) - \gamma^2 = 0 \quad ,$$

which are easily found to be

$$\nu_0 = \{1 - \frac{1}{2} [\alpha^2 + (\alpha^2 + 4\beta^2)^{1/2}]\}^{1/2} \tag{11.61}$$

$$t_0 = (1 - \gamma^2)^{1/2} \quad . \tag{11.62}$$

Here, only the physically meaningful roots are taken.

Since both $g_{m\ell}(\nu)$ and $f(t)$ are even functions of ν and t respectively , the turning points always occur in pairs $+\nu_0$ and $\pm t_0$. This means that $t(\nu)$ will be symmetric about the point $t(\nu = 0) = 0$ as mentioned already. Therefore, taking the lower integration limits in (11.58) as $-\nu_0$ on the right-hand side and $-t_0$ on the left-hand side defines the mapping of domain D of ν onto the domain D_t of t.

195

Next we consider the singularities of $g_{m\ell}(\nu)$ and $f(t)$, which are in fact the singular points of the differential equations (11.55,47). The latter has two regular singularities at ± 1 and an irregular singularity at infinity. The comparison equation (11.55) also has three singular points; however, they are all regular. Since the point at infinity is outside the domains of these functions and outside the defined physical boundaries, this difference could be considered unimportant. We have already verified that the turning points of (11.55) or (11.52,48) are mapped onto each other and consequently $t'(\nu) \neq 0$ and finite in D_ν. The remaining question then is whether $t(\nu)$ is a one-to-one mapping everywhere else. The answer is, not always; there are some cases where $[\nu_0, 1]$ is mapped onto a point.

Consider the positive interval $[0,1]$ of D_ν and D_t, and the case when $\beta^2 \neq 0$ and $\gamma^2 \neq 0$ in (11.61,62). Then, clearly $0 \leq \nu_0 \leq 1$ and $0 \leq t_0 \leq 1$, and thus the domain D_ν is mapped onto D_t as the two segments $[0,\nu_0] \to [0,t_0]$ and $[\nu_0,1] \to [t_0,1]$ and similarly for the negative interval $[-1,0]$. So far, the mapping is one-to-one. Now, let $\beta^2 = \gamma^2 = 0$. Then two cases can occur. If $\alpha^2 \leq 0$, then $\nu_0 = t_0 = 1$; again the mapping is one-to-one. When $\alpha^2 > 0$, then $\nu_0 = (1 - \alpha^2)^{1/2}$ and $t_0 = 1$. In this case the interval $[0,\nu_0]$ of D_ν is mapped entirely onto $[0,1]$ of D_t, which implies that the interval $[\nu_0,1]$ is mapped onto a point in D_t, i.e., onto $t_0 = 1$. This is a rather unwelcome feature we should like to avoid. It is, in fact, the difference in the nature of the singularities at infinity which brings about this problem.

To get out of this difficulty, we recall that the uniform WKB wave functions are semiclassical in nature and introduce the following replacements of parameters

$$\ell(\ell + 1) \to (\ell + \tfrac{1}{2})^2 \tag{11.63}$$

$$(m^2 - 1) \to (m + d)^2 \tag{11.64}$$

in (11.48,52) which together define $t'(\nu)$ in (11.53). The replacement (11.63) is the familiar Langer modification [2.19b] made for the semiclassical (WKB) radial wave functions in the spherical coordinate system to satisfy the boundary condition at $r = 0$. Except for the parameter d, (11.64) is analogous to the *Bethe* modification [11.7] for angular spheroidal wave functions which takes $d = 0$. The additional parameter d is introduced into the Bethe modification solely on the ground that for $d \neq 0$, $t(\nu)$ will be a one-to-one mapping of D_ν onto D_t when α^2, β^2, and γ^2 in (11.60) are modified according to (11.63,64). The parameter d will be more fully specified below when we discuss the calculation of the eigenvalues $\tau_{m\ell}$ in the uniform WKB method. With the above considerations given to the mapping, we now turn to the determination of $\tau_{m\ell}$ in terms of ℓ, m, and λp. That can be done by integrating (11.53):

$$\int_{-t_0}^{t_0} dt \, f^{\frac{1}{2}}(t) = \int_{-\nu_m}^{\nu_m} d\nu \, g_{m\ell}^{\frac{1}{2}}(\nu) \quad , \tag{11.65}$$

where $\pm t_0$ are the zeros of $f(t)$:

$$t_0 = (1 - \gamma^2/c^2)^{1/2} \quad .$$

Since t_0 must be real, γ must be bounded from below and above:

$$- C \leq \gamma \leq C \quad .$$

Note that this condition is consistent with the condition $- \ell \leq m \leq \ell$ mentioned earlier. The integral on the left-hand side of (11.65) is

$$\int_{-t_0}^{t_0} dt \, f^{\frac{1}{2}}(t) = \pi(C - \gamma) \quad . \tag{11.66}$$

At this point, by using (11.63,64), we find the Bohr-Sommerfeld quantization condition for periodic motion in the ν mode

$$\int_{-\nu_m}^{\nu_m} d\nu \, g_{m\ell}^{\frac{1}{2}}(\nu) = (\ell + \frac{1}{2} - m - d)\pi \quad . \tag{11.67}$$

Evaluation of the integral in (11.67) and solution of the equation will give the eigenvalues $\tau_{m\ell}$ for the angular part of the wave function $S_{m\ell}(\lambda p, \nu)$. The integral can be given generally in terms of elliptic integrals. The case of $m = 0$ and $d = 0$ is especially simple. Equation (11.67) for $m = 0$ and $d = 0$ is

$$(\ell + \frac{1}{2})\pi = \tau_{0\ell}^{\frac{1}{2}}\{E(\pi/2|c_\ell) + c_\ell^{-1}[E(\pi/2|c_\ell) - F(\pi/2|c_\ell)]\} \quad , \tag{11.68}$$

where $c_\ell = \tau_{0\ell}^{\frac{1}{2}}/\lambda p$ and F and E are respectively the complete elliptic integral of the first and second kind. This equation may be solved graphically for the eigenvalue. Numerical solution of (11.67) is quite simple. The approximations presented here for various quantities will be assessed for their accuracy in the following section.

The uniform WKB solution $y_{\ell m}(\lambda p, \nu)$ given in (11.57b) may be used for computing approximate formulas for $d_n(\lambda p | m\ell)$. This is in effect equivalent to solving the three-term recursion relation for d_n; and we may call the d_n thus obtained the WKB approximation for d_n. First, let us note that by the orthogonality of the Legendre polynomials

$$d_{\ell-m}(\lambda p|m_\ell) = \left(\frac{2\ell + 1}{2}\right)\frac{(\ell - m)!}{(\ell + m)!} \int_{-1}^{1} dv \ P_\ell^m(v)S_{m\ell}(\lambda p,v) \quad . \tag{11.69}$$

The WKB approximation for $d_{\ell-m}$ is obtained if $S_{m\ell}$ is replaced with $y_{\ell m}$ in (11.69):

$$d_{\ell-m}(\lambda p|m\ell) = \left(\frac{2\ell + 1}{2}\right)\frac{(\ell - m)!}{(\ell + m)!} \int_{-1}^{1} dv \ P_\ell^m(v)y_{\ell m}(\lambda p,v) \quad . \tag{11.70}$$

11.5 Accuracy of the Uniform WKB Approximation

We now evaluate the accuracy of the uniform WKB approximation to the angular spheroidal wave functions presented in the previous section by direct numerical comparison. This will simultaneously enable us to consider practical methods of computing various quantities developed so far.

11.5.1 Eigenvalues

To calculate either of the angular spheroidal wave functions $S_{m\ell}(\lambda p,v)$ or $y_{m\ell}(v)$, we first require the eigenvalues $\tau_{m\ell}$, which may be computed from the semiclassical formula (11.67). Given a set of parameters λp, ℓ, and m which determine the turning points v_0 and ψ_0, the root of the function

$$\Theta(v_0,\psi_0,\tau_{m\ell}) = \int_{-v_0}^{v_0} dy \ g_{m\ell}(y) - (\ell + \tfrac{1}{2} - m - d) \quad ,$$

namely, the eigenvalue, can be found by Newton's method quite easily. To implement the actual computation, we must specify the parameter d. For small values of the parameter λp and for a wide range of d values, *Sink* and *Eu* [11.8] compared the semiclassical eigenvalues with those obtained by solving the recursion relation for d_n and found that if the parameter is chosen to be

$$d = 1/3\lambda p \quad , \tag{11.71}$$

the eigenvalues for small ℓ and m become quite accurate. Tables 11.1,2 compare the semiclassical eigenvalues $\tau_{m\ell}(d)$ with the exact values for $\tau_{m\ell}$ when d = 0 and d = 1/3λp for λp = 1 and 8, respectively. These results show that the choice of (11.71) improves the low-quantum-number eigenvalues quite significantly, while the effect for larger ℓ is quite small. The modification in (11.64) was introduced to insure the one-to-one mapping for the function t(v), but the numerical results for the eigenvalues also show that there is a rather significant practical implication, since it improves the eigenvalues

Table 11.1. Comparison of semiclassical and exact eigenvalues for
$\lambda p = 1$, $m = 0$

	Eigenvalues ($\tau_{m\ell}$)		
ℓ	$d = 0$	$d = 1/3$	Exact
0	0.8503	0.4532	0.3190
1	2.8173	2.7345	2.5931
2	6.9164	6.7452	6.5335
3	13.0757	12.7463	12.5145
4	21.2901	20.7441	20.5083
5	31.5590	30.7379	30.5054
6	43.8818	42.7258	42.5038
7	58.2587	56.7059	56.5028
8	74.6895	72.6771	72.5022
9	93.1742	90.6394	90.5017

Table 11.2. Comparison of semiclassical and exact eigenvalues for $\lambda p = 8$

		Eigenvalues ($\tau_{m\ell}$)		
ℓ	m	$d = 0$	$d = 1/24$	Exact
0	0	7.8728	7.2285	7.2216
1		22.8147	22.2171	22.0921
2		36.5650	36.0212	35.7064
3		48.9019	48.4249	47.7571
4		59.3670	58.9903	58.0167
5		67.2125	66.9799	67.3646
6		78.0284	77.5197	77.8251
7		91.5871	90.8551	90.6913
8		107.4998	106.5508	106.0116
9		125.6305	124.4599	123.5771
1	1	8.9415	8.3794	8.3000
2	2	12.1432	11.6511	11.5278
3	3	17.4671	17.0335	16.8866
4	4	24.8978	24.5129	24.3548
5	5	34.4189	34.0747	33.9121
6	6	46.0151	45.7048	45.5411
7	7	59.6730	59.3912	59.2276
8	8	75.3813	75.1230	74.9605
9	9	93.1308	92.8938	92.7310

for low ℓ. Clearly, (11.71) is valid only for $3\lambda p > 1$. The case of $3\lambda p < 1$
does not matter, since the semiclassical method then would not be valid any-
way. In such cases a numerical solution method must be used to calculate
eigenvalues.

Once the eigenvalues $\tau_{m\ell}(d)$ are known, it is simple to obtain the function
$t(\nu)$ which maps D_ν to D_t. The method is identical with that employed to obtain

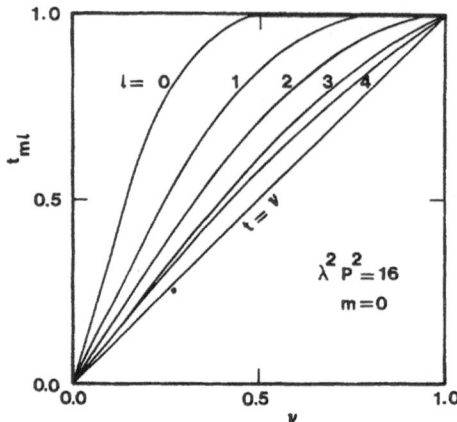

Fig.11.5. The transformation (mapping) function $t(\nu)$ versus ν

the eigenvalues. For a set of parameters λp, m, ℓ, $\tau_{m\ell}(d)$, and ν, the corresponding value for t is found as the root of the function $\Theta(\nu,\psi,\tau_{m\ell})$. The behavior of $t(\nu)$ is shown in Fig.11.5 for the specific case of $\lambda p = 4$ and $m = 0$. The mapping $t(\nu)$ for $\ell = 0$ is highly nonlinear; however, as ℓ increases, $t(\nu)$ rapidly approaches the identity mapping $t = \nu$. This result is quite interesting as it shows that the angular spheroidal wave functions begin to resemble the associated Legendre polynomials $P_\ell^m(\nu)$ as ℓ increases, as expected, but much faster than anticipated. This behavior can be taken advantage of when asymptotic forms are developed for $y_{m\ell}(\mu)$ for large ℓ.

11.5.2 Angular Spheroidal Wave Functions

Now that the mapping function $t(\nu)$ is determined, the uniform WKB approximation $y_{m\ell}(\nu)$ for the angular spheroidal wave function can be calculated from (11.57b). In the actual calculation the associated Legendre polynomials $P_\ell^m(t(\nu))$ are computed by means of a recurrence relation and the normalization $N_{m\ell}$ is computed according to (11.57c). Some examples of angular spheroidal functions $y_{m\ell}$ thus calculated are shown in Fig.11.6 for $\lambda p = 4$ and $\ell = 0 - 3$ along with the angular spheroidal wave functions $S_{m\ell}(\lambda p,\nu)$ obtained by (11.7). The expansion coefficients d_n were taken from the tables of *Stratton* et al. [11.6], and the wave functions then normalized to unity according to (11.8). For $\ell = 0$, the agreement is qualitative, but the accuracy of the uniform WKB wave functions increases dramatically. By the time ℓ is 3, the agreement is quantitative over the entire domain. The error in $y_{m\ell}(\nu)$ for $\ell \geq 3$ is generally much less than 1% and as ℓ increases, the uniform WKB wave functions become excellent in accuracy [11.8].

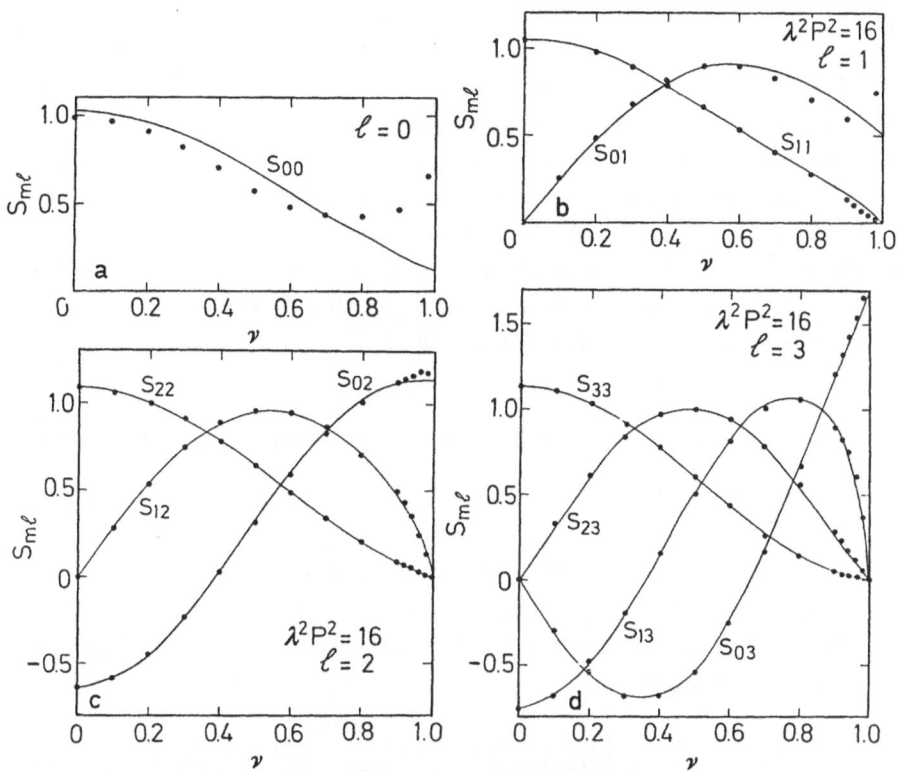

Fig.11.6a-d. Approximate and exact angular spheroidal wave functions for
$\ell = 0,1,2,3$ at $\lambda p = 4$. The dots are the approximate wave functions $Y_{m\ell}$ and
the solid lines are the exact wave functions $S_{m\ell}$

As ℓ and m increase, the associated Legendre polynomials become difficult
to calculate accurately with a recurrence relation due to the propagation of
truncation errors. In such cases we need a different procedure which is indicated below.

In the previous subsection it is shown that the mapping function $t(\nu)$
tends to approach the identity $t = \nu$ as ℓ increases. Sink and Eu [11.8] found
that the error in using the identity transformation $t = \nu$ is less than 1%
when $|\tau_{0L}|$ increases to approximately $5\lambda^2 p^2$. Thus for $\ell > L$ the function
$t(\nu)$ may be simply replaced with ν itself. When ℓ is large, the eigenvalues
are approximately equal to

$$\tau_{0\ell} \approx \ell(\ell + 1) + \lambda^2 p^2/2 \quad . \tag{11.72}$$

With the above rule of thumb $\tau_{0L} \approx 5\lambda^2 p^2$ for L, we find the approximate
value for L

$$L \approx 3\lambda p/\sqrt{2} - \frac{1}{2} \; . \tag{11.73}$$

For example, $L \approx 8$ for $\lambda p = 4$, and $L \approx 135$ for $\lambda p = 64$. Since calculating $t(\nu)$ is generally the most time consuming part of evaluating $Y_{m\ell}(\nu)$, a large reduction in computational effort can be realized by this additional approximation. In this connection we note that $\nu(t)$ can be expressed in terms of various elliptic integrals, which then must be inverted to find $t(\nu)$ as a function of ν. This step cannot be done analytically in general, and what is given above is an approximate procedure for this inversion.

When $t(\nu) = \nu$, then (11.57b) reduces to

$$Y_{m\ell}(\nu) = N_{m\ell} P_{\ell}^{m}(\nu) \; . \tag{11.74}$$

Using the normalization of the associated Legendre polynomials, we then find

$$N_{m\ell}^{2} = \frac{(2\ell + 1)}{2} \frac{(\ell - m)!}{(\ell + m)!} \; . \tag{11.75}$$

Because a large number of these angular functions will be required for scattering problem (e.g., $\ell \sim 100$), further steps can be taken to simplify the calculation of $Y_{m\ell}(\nu)$ for $\ell > L$. From (11.74), we can derive the recurrence relation for $Y_{m\ell}$:

$$(\ell + 1 - m)Y_{m,\ell+1}(\nu) = \left(\frac{(2\ell + 3)(\ell + 1 - m)}{(2\ell + 1)(\ell + 1 + m)}\right)^{1/2} (2\ell + 1)Y_{m,\ell}(\nu)$$

$$- \left(\frac{(2\ell + 3)(\ell + 1 - m)(\ell - m)}{(2\ell - 1)(\ell + 1 + m)(\ell + m)}\right)^{1/2} (\ell + m)Y_{m,\ell-1}(\nu) \tag{11.76}$$

and

$$Y_{m+2,\ell}(\nu) = [(\ell + m + 2)(\ell - m - 1)]^{-1/2} \frac{2(m+1)\nu}{\sqrt{1 - \nu^2}} Y_{m+1,\ell}(\nu)$$

$$- [(\ell + m + 2)(\ell + m + 1)(\ell - m)(\ell - m - 1)]^{-1/2}$$

$$\times (\ell - m)(\ell + m + 1)Y_{m,\ell}(\nu) \tag{11.77}$$

for concurrent ℓ and m values. These recurrence relations are for the normalized asymptotic angular functions of (11.74). Thus the normalization step may be disposed of, too. In practice, (11.76) has been found to be more useful than (11.77) and it is found [11.8] that little round-off error is accumulated if (11.76) is recast into the form

$$Y_{m,\ell+1}(\nu) = A[2\ell\nu Y_{m,\ell}(\nu) - \ell B Y_{m,\ell-1}(\nu)]$$

$$+ \nu Y_{m,\ell} - m B Y_{m,\ell-1} \; , \tag{11.78}$$

where

$$A = [(2\ell + 3)/(\ell + 1 - m)(\ell + 1 + m)(2\ell + 1)]^{1/2}$$

$$B = [(\ell - m)(2\ell + 1)/(2\ell - 1)(\ell + m)]^{1/2} .$$

The recurrence relation (11.78) for ascending ℓ then requires only the key function $Y_{m,m}(\nu)$, whereupon

$$Y_{m,m+1}(\nu) = (2m + 3)^{1/2} \nu Y_{mm} \tag{11.79}$$

and $Y_{m\ell}(\nu)$ for $\ell \geq m + 2$ then may be obtained from (11.78). The key function $Y_{mm}(\nu)$ can be calculated to a good approximation by the following asymptotic expansion:

$$Y_{mm}(\nu) = [(2m + 1)/(4\pi m)^{1/2}]^{1/2}(1 - \nu^2)^{m/2} \exp[K(m)] , \tag{11.80}$$

where $K(m)$ is an asymptotic series

$$K(m) = \sum_{r=1}^{\infty} \frac{(-)^{r-1}B_r}{2r(2r - 1)m^{2r-1}} (2^{-2r} - 1) . \tag{11.81}$$

The B_r are the Bernoulli numbers. The asymptotic expansion for $Y_{m\ell}$ above can be easily derived from (11.74) which is a valid approximation for large ℓ. When Rodrigue's formula is used for $P_\ell^m(t)$, we obtain

$$Y_{mm}(\nu) = \left(\frac{(2m + 1)}{2} \frac{1}{(2m)!}\right)^{1/2}(1 - \nu^2)^{m/2} \frac{d^m}{d\nu^m} P_m(\nu) . \tag{11.82}$$

Since

$$\frac{d^m}{d\nu^m} P_m(\nu) = \frac{(2m)!}{2^m m!} , \tag{11.83}$$

from (11.79) it follows

$$Y_{mm}(\nu) = \left(\frac{2m + 1}{2}\right)^{1/2} \frac{[(2m)!]^{\frac{1}{2}}}{2^m m!} (1 - \nu^2)^{m/2} . \tag{11.84}$$

This expression is simple enough to be practical for calculating Y_{mm} for $m \leq 40$. For $m \geq 40$, overflow begins to be a problem in the numerical computation. To avoid this problem, a more versatile asymptotic expression can be derived by using the Sterling series

$$\ln\Gamma(z) = (z - \tfrac{1}{2})\ln(z) - z + \tfrac{1}{2} \ln(2\pi) + \gamma(z) \tag{11.85}$$

for the gamma function. The term $\gamma(z)$ in (11.85) is the asymptotic Sterling series

$$\gamma(z) = \sum_{r=1}^{\infty} (-)^{r-1} B_r / 2r(2r - 1) z^{2r-1} \quad , \tag{11.86}$$

where B_r are the Bernoulli numbers. Using (11.85), we find that

$$\frac{[(2m)!\,]^{1/2}}{2^m m!} = (\pi m)^{-1/4} \exp[\gamma(2m)/2 - \gamma(m)] \quad . \tag{11.87}$$

We then define $K(m) = \gamma(2m)/2 - \gamma(m)$ which can easily be shown to have the same form as (11.81), when (11.86) is used. This completes the derivation of the asymptotic formula (11.80).

Being asymptotic, the series (11.81) does not converge, but the first few terms give results of excellent accuracy. According to the calculation performed by *Sink* [11.9], the first seven-term approximation of (11.8) yields $y_{10,10}(0)$ to fifteen significant digits of accuracy, while the same accuracy for $y_{100,100}(0)$ is found by using only the first three terms.

11.5.3 Elastic Cross Section

It is now relatively simple to calculate differential elastic-scattering cross sections (11.25) by using the WKB phase-shift formula in Sect.11.2 and the approximation procedure for $S_{m\ell}$ described in Sect.11.4 and in the previous subsections. Since the scattering amplitudes depend on the direction of the incident beam relative to the major axis of the ellipsoid, the calculation is a little involved, but still much simpler than the calculation in the spherical coordinate representation which requires the solution of close-coupled Schrödinger equations [11.10]. In Fig.11.7 some differential elastic cross sections are presented for three different values of λp [11.11]. They are for a hard ellipsoid and look similar to the elastic differential cross sections for the Li^+ - H_2 system calculated by *Schinke* [11.10]. The comparison is only qualitative, probably due to the assumption of a hard ellipsoid and the difference in energy.

11.6 Inelastic Scattering

The foregoing theory can be applied to studying inelastic scattering of a particle by a hard ellipsoid with attractive potential tail. For the purpose we assume that the potential has the form

$$V = \sum_{ab} V_{ab}(\mu,\nu) Y(ba|q,\nu,\phi) D_{b0}^a(\Omega) \quad , \tag{11.88}$$

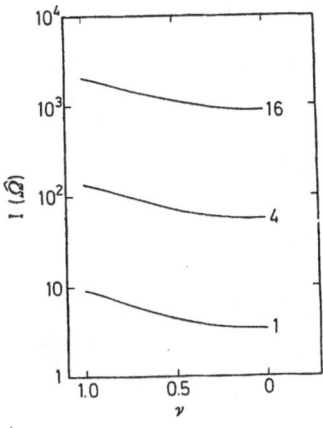

Fig.11.7. Differential elastic cross sections at different values of λp

where

$$Y(ba|q,\nu,\phi) = S_{ba}(q,\nu)\Phi_b(\phi) \tag{11.89}$$

$$V_{ab}(\mu,\nu) = V_{ab}(\mu)/(\mu^2 - \nu^2) \quad , \tag{11.90}$$

with $V_{ab}(\mu)$ satisfying the same conditions as $V(\mu)$ in (11.41b). We may assume for $V_{ab}(\mu)$ the same mathematical form as in (11.41c). Since the ellipsoid is hard at $\mu = \mu_0 > 1$, then

$$\lim_{\mu \to \mu_0} V_{ab}(\mu) = \infty \quad . \tag{11.91}$$

The q in (11.89) is an adjustable parameter pertinent to the angular dependence of the potential. Since $Y(ba|q,\nu,\phi)$ tends to the spherical harmonic $Y_a^b(\nu,\phi)$ apart from a constant normalization factor as $q \to 0$ (Sect.11.5.2), the potential function assumed in (11.88) is a prolate spheroidal coordinate representation analog of the potential function,

$$V = \sum_{ab} \sum_c V_{ab}(R)Y_a^{c*}(\theta,\psi)D_{cb}^a(\Omega) \quad , \tag{11.92}$$

generally assumed in the spherical coordinate representation [3.14,15]. The wave function for an ellipsoid may be written as

$$Y(LKM_L|\Omega) = [(2L + 1)/8\pi^2]^{1/2}D_{-K-M_L}^L(\Omega) \quad , \tag{11.93}$$

where L,K, and M_L are the angular momentum of the body, its projections on the body-fixed and space-fixed axis, respectively.

Then writing the entire wave function for the system in the form

$$\Psi(LKM_L m\ell | R,\Omega) = \sum_{LKM_L m\ell} R(LKM_L m\ell;\mu) Y(m\ell | \lambda p,\nu,\phi) V(LKM_L | \Omega) \quad , \tag{11.94}$$

and substituting it into the Schrödinger equation, it is possible to derive a set of coupled Schrödinger equations for $R(LKM_L m\ell;\mu)$:

$$\left[\frac{d}{d\mu}(\mu^2 - 1)\frac{d}{d\mu} + \lambda^2 p^2 \mu^2 - \tau_{m\ell} - \frac{m^2}{\mu^2 - 1}\right] R(LKM_L m\ell;\mu)$$

$$= \lambda^2 \sum_{ab} \sum_{L'K'M_L'm'\ell'} V_{ab}(\mu) Z_{ab}(LKM_L m\ell | L'K'M_L'm'\ell') R(L'K'M_L'm'\ell';\mu) \quad ,$$

where

$$\tag{11.95}$$

$$Z_{ab}(LKM_L m\ell | L'K'M_L'm'\ell') = \int d\nu \int d\phi \int d\Omega \; Y^*(m\ell | \lambda p,\nu,\phi) V^*(LKM_L | \Omega)$$

$$\times \; Y(ba | q,\nu,\phi) \; D_{b0}^a(\Omega) \; Y(m'\ell' | \lambda p,\nu,\phi) \; V(L'K'M_L | \Omega) \; .$$

$$\tag{11.96}$$

Equation (11.96) is equivalent to the close-coupled equations in the spherical coordinate representation theory. An advantage of (11.96) is that the radial part of the wave functions have an already built-in nonspherical geometry that has an advantage over the spherical coordinate representation for an inherently nonspherical system. The semiclassical methods developed for coupled radial Schrödinger equations in the previous chapters, of course, apply to (11.95). It would be repetitive to develop such approximation theories again here.

Equation (11.95), however, is not in the most ideal form for studying inelastic transitions among the rotational states, because the irreducible representation of the rotation group is not employed. It is clear that the rotation group for an ellipsoid is a subgroup of the full rotation group for a sphere, and one may expand the wave functions accordingly, thereby simplifying the coupled equations. This part of the problem awaits further careful study, which would be undoubtedly worthwhile and even important.

12. Concluding Remarks

This monograph represents a study of a class of semiclassical methods for analysing atomic and molecular scattering which appears more systematic than some ad hoc approximation methods often termed semiclassical. The latter are, in fact, quite wide ranging. The subjects chosen reflect the author's past work and attitude towards approximations in molecular scattering theory. Our scientific tastes are varied, and one man's delight may turn out to be another's indifference. In any event, despite the enhanced accuracy experimental techniques have achieved, approximation methods are not only useful and practical, but also a necessity, although immense progress has been made in numerical methods with the help of digital computers. Judicious combination of approximations and numerical methods appears to hold out promise, and an important point of the present work is in advocating that notion towards collision problems. As shown with some examples, the WKB method, when treated correctly, is probably the most powerful and accurate among various approximation methods at the lowest level of approximation. Since systematic approximation schemes are not realizable divorced from some observance of mathematical rigor, attempts have been made to preserve the rigor wherever possible and practicable.

There is probably little of major importance left in the WKB approximation for elastic scattering, although it is pedagogically useful, giving much insight into more important problems such as inelastic and reactive scattering.

The uniform WKB theory rests on the idea of getting for complicated, unsolvable potentials the renormalized wave functions — if we may use a currently fashionable term "renormalization" — which look like the wave functions for similar, but easily solvable potentials. I believe the terminology is apt, since the theory involves essentially renormalization transformations which put an unsolvable problem into a solvable *normal* form asymptotically. The idea is fertile and presents many possibilities of application to molecular problems otherwise not amenable to analytic solutions.

In the uniform WKB theory first-order Magnus approximations or variations thereof have often been used to construct the S-matrix elements which remain unitary regardless of the approximation used. The first-order Magnus approximation is a kin to the so-called infinite-order sudden approximation, which is in turn mathematically rather closely related to the first-order cumulant approximation to the S matrix. Their mutual connections have not been studied in this work, but clarification of them would be intrinsically interesting and could provide some practical ways to enhance their precision by grafting one method onto another.

The time-dependent semiclassical theory and the semiclassical theory of multisurface systems are largely untapped of their potential. The problems involved are difficult and the present work has barely scratched the surface. It would take some hard-nosed mathematical analysis before they are made sufficiently accurate and practical. In semiclassical theories a missing small phase factor in oscillatory terms often causes great damage to the accuracy. One must probably be mindful of such factors in the time-dependent semiclassical theories as well and look for them, if the results do not seem to work out numerically as expected. It can be quite frustrating to try to find small phase factors, but it is also quite possible that one will be amply compensated for the pain, if successful, since the time-dependent semiclassical theory appears basically sound.

The formal results obtained for the S matrix in Chaps.6 and 10 point to some interesting possible schemes whereby one can combine classical trajectory methods with quantum-mechanical momentum distribution functions of the initial and final internal states. Since the momentum distribution functions represent ensembles of systems distributed over the internal states at a given energy, the classical evolution given by the classical action integral may be viewed as a transformation of the initial to the final ensemble.

Curve-crossing is one model of collision phenomena that has much benefited by employing the semiclassical theory. It renders qualitatively correct results under some simplifying assumptions, such as those employed in the Landau-Zener-Stückelberg theory, much improving our understanding of inelastic processes. The belief that the same would hold true with multichannel problems was the principal motivation for our studies of curve-crossing problems of two or more channels. As the number of channels increases, the model poses mathematical problems, an aspect of which was studied in Chaps.8 and 9. In these studies qualitative, at least quasianalytic, features of the theoretical results have been emphasized, since they can be much more insightful than a set of more accurate, but limited numerical results. Of course, removing some of the assumptions made in the study would be useful and could form an interesting subject of study. The problem is mathematically quite reminiscent of the nearest-neighbor interaction problems in the many-body theory and one may be able to dream up some intriguing theories.

The semiclassical theory of ellipsoidal-particle scattering stems from a desire to pay more attention to the geometric reality of molecules when developing approximation methods. It seems that one could enhance the accuracy of an approximation even at the lowest order, if the molecular geometry is respected as much as possible. This concept gets more credible, especially as any formal series in, say, spherical coordinates for the potential function for a molecule of nonspherical geometry is mutilated to enhance practicability in numerical computation, thereby the representation rendered physically unrealistic. The theory has potential significance for inelastic scattering and particularly for rotational inelastic scattering, but the latter is not sufficiently developed as yet. Since due to the loss of a symmetry element in the case of an ellipsoid we no longer have the full rotational group to exploit in simplifying the equations, the close-coupled equations become more complicated and larger in number. Reformulating the theory is not very difficult and the coupled equations presented clearly indicate the way to be taken. Nevertheless, the question is open at the present time. In any event, the approximation methods developed in the preceding chapters are applicable to the close-coupled equations obtained in elliptic coordinates.

I suppose one embarks on scientific endeavors with many hopes of nice things achieved at the end, but, alas, one never seems able to escape the feeling that the goal has not been achieved to his satisfaction. The present work is no exception and there are many problems left either completely untouched or unsolved, conveniently left for future study. One needs time to look back and reflect on what is done and is available. It is hoped that this work will in some small measure stimulate further interests in semiclassical methods.

Appendix 1. Asymptotic Forms for Parabolic Cylinder Functions

As $|z| \rightarrow \infty$,

$$D_\nu(z) \rightarrow z^\nu \exp(-z^2/4) \quad (-3\pi/4 < \arg z < 3\pi/4)$$

$$D_\nu(z) \rightarrow z^\nu \exp(-z^2/4) - \frac{(2\pi)^{1/2}}{\Gamma(-\nu)} \exp(\pi\nu i)z^{-\nu-1} \exp(z^2/4)$$

$$(\pi/4 < \arg z < 5\pi/4)$$

$$D_\nu(z) \rightarrow z^\nu \exp(-z^2/4) - \frac{(2\pi)^{1/2}}{\Gamma(-\nu)} \exp(-\pi\nu i)z^{-\nu-1} \exp(z^2/4)$$

$$(-5\pi/4 < \arg z < -\pi/4) \quad .$$

When $x \gg |a|$,

$$W(a;x) \rightarrow (2\kappa/x)^{1/2} \cos(x^2/4 - a \log x + \pi/4 + \psi/2)$$

$$W(a;-x) \rightarrow (2/\kappa x)^{1/2} \sin(x^2/4 - a \log x + \pi/4 + \psi/2)$$

$$\psi = \arg\Gamma(1/2 + ia)$$

$$\kappa = [1 + \exp(2\pi a)]^{1/2} - \exp(\pi a) \quad .$$

Commonly Used Symbols

The numbers after the definition of a symbol are the page numbers where the symbols are defined.

$Ai(\pm z)$	regular Airy function 10,12,13,38,99
$Bi(\pm z)$	irregular Airy function 10,13,38
$A(r)$	regular uniform WKB wave function 38
$B(r)$	irregular uniform WKB wave function 38
A_j	regular uniform WKB wave function for channel j 59,61,65, 70,75,129,147
B_j	irregular uniform WKB wave function for channel j 59,61, 62,65,70,75,129,147
(A_i,A_j)	the Wronskian of A_i and A_j. The Wronskians may be similarly formed with different pairs of uniform WKB wave functions 62,115,116,118
$I(Z_i,Z_j)_v$	the integral of Wronskian (Z_i,Z_j) from v_t to v 112
$D_n(\pm t)$	parabolic cylinder function 45,147
$W(\alpha;\pm t)$	parabolic cylinder function 46,47
$E_i(r)$	energy minus the diagonal effective potential 3
$E(\pi/2\|\gamma_\ell)$	elliptic integral of the first kind 188
$F(\phi\|c_\ell)$	elliptic integral of the second kind 189
$G^{(\pm)}$	Green's function 18,19,23
$G_0^{(\pm)}$	free Green's function 18,19
$G_a^{(\pm)}$	free Green's function for channel a 22
H	Hamiltonian 17,18,22,166

References

Chapter 1

1.1 E.H. Taylor, S. Datz: J. Chem. Phys. **23**, 1711 (1955)
1.2 H. Pauly, J.P. Toennies: Adv. At. Mol. Phys. **1**, 201 (1965)
     J. Ross (ed.): Adv. Chem. Phys. **10** (1966)
     J.P. Toennies: Appl. Phys. **3**, 91 (1974)
1.3 R.E. Smalley, B.L. Ramakrishna, D.H. Levy, L. Wharton: J. Chem. Phys.
     61, 4365 (1974)
1.4 E.R. Grant, P.A. Schulz, Aa.S. Sudbø, Y.R. Shen, Y.T. Lee: Phys. Rev.
     Lett. **40**, 115 (1978)
1.5 J. Liouville: J. Math. Pures Appl. **2**, 16 (1837)
1.6 C.G. Callan, Jr., R.F. Dashen, D.J. Gross: AIP Conf. Proc. **55** (1979)
1.7 G.G. Stokes: Trans. Cambridge Philos. Soc. **9**, 165 (1856)

Chapter 2

2.1 H. Poincaré: Acta Math. VIII, 295 (1886)
2.2 E.T. Whittaker, G.N. Watson: *Modern Analysis*, 4th ed. (Cambridge University Press, London 1952)
2.3 G.G. Stokes: Trans. Cambridge Philos. Soc. **9**, 165 (1856); **10**, 105 (1857); **11**, 412 (1871)
2.4 P.M. Morse, H. Feshbach: *Methods of Theoretical Physics* (McGraw-Hill, New York 1953)
2.5 H. Jeffreys: *Asymptotic Approximations* (Oxford University Press, London 1962)
2.6 A. Erdelyi: *Asymptotic Expansions* (Dover, New York 1956)
2.7 J. Liouville: J. Math. Pures Appl. **2**, 16 (1837)
2.8 J. Horn: Math. Ann. **52**, 271 (1899)
2.9 Lord Rayleigh: Proc. R. Soc. London, Ser. A**86**, 207 (1912)
2.10 G. Wentzel: Z. Phys. **38**, 518 (1926)
2.11 H.A. Kramers: Z. Phys. **39**, 828 (1926)
2.12 M.L. Brillouin: J. Phys. Radium **7**, 353 (1926)
2.13 G.N. Watson: *Theory of Bessel Functions* (Cambridge University Press, London 1966)
2.14 M. Abramowitz, I.A. Stegun: *Handbook of Mathematical Functions* (NBS, Washington, DC 1964)
2.15 J. Heading: *An Introduction to Phase Integral Methods* (Methuen, London 1962)
2.16 A. Zwaan: Intensitäten im Ca-Funkenspectrum, Ph.D. Dissertation, Utrecht (1929)
2.17 E.C. Kemble: *The Fundamental Principles of Quantum Mechanics* (McGraw-Hill, New York 1937)
2.18 R.E. Langer: Bull. Amer. Math. Soc. (2) **40**, 545 (1934); Phys. Rev. **51**, 669 (1937); Commun. Pure Appl. Math. **3**, 427 (1950); Trans. Am. Math. Soc. **84**, 144 (1957)

Chapter 3

3.1 M.L. Goldberger, K.M. Watson: *Collision Theory* (Wiley, New York 1964)
3.2 N.F. Mott, H.S.W. Massey: *The Theory of Atomic Collisions* (Oxford University Press, London 1965)
3.3 R.G. Newton: *Scattering Theory of Waves and Particles* (McGraw-Hill, New York 1966)
3.4 R.B. Bernstein (ed.): *Atom-Molecule Collision Theory* (Plenum, New York 1979)
3.5 E. Prugovecki: *Quantum Mechanics in Hilbert Space* (Academic, New York 1971)
3.6 R.D. Levine: *Quantum Mechanics of Molecular Rate Processes* (Clarendon Press, Oxford 1969)
3.7 P.M. Morse, H. Feshbach: *Methods of Theoretical Physics* (McGraw-Hill, New York 1953)
3.8 F.B. Hildebrand: *Methods of Applied Mathematics* (Prentice-Hall, Englewood Cliffs NY 1961)
3.9 W. Haack, W. Wendland: *Lectures on Partial and Pfaffian Differential Equations* (Pergamon, New York 1972)
3.10 E.N. Economou: *Green's Functions in Quantum Physics*, 2nd ed., Springer Ser. Solid-State Sci., Vol. 7 (Springer, Berlin, Heidelberg, New York 1983)
3.11 C. Møller: K. Dan. Vidensk. Selsk. Mat.-Fys. Medd. **23** (1) (1945)
3.12 H. Ekstein: Phys. Rev. **101**, 880 (1956)
3.13 L.D. Faddeev: *Mathematical Aspects of the Three-body Problem in the Quantum Scattering Theory* (Daniel Davey, New York 1965)
3.14 K. Takayanagi: Prog. Theor. Phys. Suppl. **25**, 1 (1963)
3.15 D. Secrest: In *Molecular Collision Dynamics*, ed. by J.M. Bowman, Topics Curr. Phys., Vol. 33 (Springer, Berlin, Heidelberg, New York 1983)
3.16 E.P. Wigner, L. Eisenbud: Phys. Rev. **72**, 29 (1947)
3.17 E.P. Wigner: *Group Theory* (Academic, New York 1959)
3.18 A.R. Edmonds: *Angular Momentum in Quantum Mechanics* (Princeton University Press, Princeton 1957)
3.19 M.E. Rose: *Elementary Theory of Angular Momentum* (Wiley, New York 1957)
3.20 E.U. Condon, G.H. Shortley: *The Theory of Atomic Spectra* (Cambridge University Press, London 1935)
3.21 I.I. Sobelman: *Atomic Spectra and Radiative Transition*, Springer Ser. Chem. Phys., Vol. 1 (Springer, Berlin, Heidelberg, New York 1979)
3.22 J.M. Blatt, L.C. Biedenharn: Rev. Mod. Phys. **24**, 258 (1952)
3.23 W.A. Lester, Jr.: Methods Comput. Phys. **10**, 211 (1971)
3.24 D. Secrest: Methods Comput. Phys. **10**, 243 (1971)
3.25 B.R. Johnson: J. Comput. Phys. **13**, 445 (1973)
3.26 Proc. Workshop on Algorithms and Computer Codes for Atomic and Molecular Quantum Scattering Theory, NRCC Proceedings No. 5, Lawrence Berkeley Laboratory, University of California (1980)

Chapter 4

4.1 W.H. Furry: Phys. Rev. **71**, 360 (1947)
4.2 C.E. Hecht, J.E. Mayer: Phys. Rev. **106**, 1156 (1957)
4.3 K.W. Ford, D.L. Hill, M. Wakano, J.A. Wheeler: Ann. Phys. (NY) **7**, 239 (1959)
4.4 K.W. Ford, J.A. Wheeler: Ann. Phys. (NY) **7**, 259 (1959)
4.5 H. Moriguchi: J. Phys. Soc. Jpn. **14**, 1771 (1959); Prog. Theor. Phys. **23**, 750 (1960)
4.6 S.I. Choi, J. Ross: Proc. Nat. Acad. Sci. (USA) **48**, 803 (1962)

4.7 C.L. Beckel, J. Nakhleh: J. Chem. Phys. **39**, 94 (1963)
4.8 M. Rosen, D.R. Yennie: J. Math. Phys. **5**, 1505 (1964)
4.9 N. Fröman, P.O. Fröman: *JWKB Approximation* (North-Holland, Amsterdam 1965)
4.10 N.I. Zhirnov: Izv. Vyssh. Uchebn. Zaved. Fiz. **12**, 86 (1969)
4.11 M.V. Berry, K.E. Mount: Rep. Prog. Phys. **35**, 315 (1972)
4.12 B.C. Eu, H. Guerin: Can. J. Phys. **52**, 1805 (1974)
4.13 G.E. Zahr, W.H. Miller: Mol. Phys. **30**, 951 (1975)
4.14 L.S. Vasilevsky, N.I. Zhirnov: J. Phys. B**10**, 1425 (1977)
4.15 H.J. Korsch, R. Möhlenkamp: J. Phys. B**10**, 3451 (1977)
4.16 B.J.B. Crowley: J. Phys. A**11**, 509 (1978); **13**, 1227 (1980)
4.17 J.N.L. Connor, H.R. Mayne: Mol. Phys. **37**, 1 (1979)
4.18 C.K. Chan, P. Lu: J. Chem. Phys. **74**, 3625 (1981)
4.19 R.B. Gerber: Mol. Phys. **42**, 693 (1981)
4.20 R.B. Bernstein: Adv. Chem. Phys. **10**, 75 (1966)
4.21 E.F. Greene, A.L. Moursund, J. Ross: Adv. Chem. Phys. **10**, 135 (1966)
4.22 H.S.W. Massey, C.B.O. Mohr: Proc. R. Soc. London Ser. A**144**, 188 (1934)
4.23 J.L. Dunham: Phys. Rev. **41**, 713 (1932)
4.24 F.W.J. Olver: Philos. Trans. R. Soc. London Ser. A**250**, 479 (1958)
4.25 A. Erdelyi: J. Math. Phys. **1**, 16 (1960)
4.26 V. Volterra, B. Hostinsky: *Operations Infinitesimales Lineaires* (Gauthier-Villars, Paris 1938)
4.27 J.D. Dollard, C.N. Friedman: *Product-Integration* (Addison-Welsey, Reading, MA 1979)
4.28 C. DeWitt-Morette, A. Maheshwari, B. Nelson: Phys. Rep. **50**, 255 (1979)
4.29 W. Magnus: Commun. Pure Appl. Math. **7**, 649 (1954)
4.30 D.W. Robinson: Helv. Phys. Acta **36**, 140 (1963)
4.31 P. Pechukas, J.C. Light: J. Chem. Phys. **44**, 3897 (1966)
4.32 R.D. Levine: Mol. Phys. **22**, 497 (1971)
4.33 R.B. Bernstein: J. Chem. Phys. **33**, 795 (1960)
4.34 W.H. Miller: J. Chem. Phys. **48**, 1651 (1981)
4.35 S.C. Miller, Jr., R.H. Good, Jr.: Phys. Rev. **91**, 174 (1953)
4.36 N.D. Kazarinoff: Arch. Ration Mech. Anal. **2**, 129 (1958)
4.37 A. Erdelyi, W. Magnus, F. Oberhettinger, E.G. Tricomi: *Higher Transcendental Functions*, Vol. 2 (McGraw-Hill, New York 1953)
4.38 L.D. Landau, E.M. Lifshitz: *Quantum Mechanics* (Pergamon, London 1958)
4.39 M.S. Child: *Molecular Collision Theory* (Academic, New York 1974)

Chapter 5

5.1 G.J. Kynch: Proc. Phys. Soc. London Ser. A**65**, 83 (1952)
5.2 G. Zemach: Nuovo Cim. **33**, 939 (1964)
5.3 A. Degasperis: Nuovo Cim. **34**, 1667 (1964)
5.4 J.R. Cox: Nuovo Cim. **37**, 474 (1965
5.5 F. Calogero: *Variable Phase Approach to Potential Scattering* (Academic, New York 1967)
5.6 C.F. Curtiss: J. Chem. Phys. **52**, 4832 (1970)
5.7 C.F. Curtiss: J. Chem. Phys. **71**, 1150 (1979)
5.8 N.G. Van Kampen: Physica **25**, 70 (1967)
5.9 B.C. Eu: J. Chem. Phys. **52**, 1882 (1970)
5.10 B.C. Eu: J. Chem. Phys. **52**, 3903 (1970)
5.11 E.C.G. Stückelberg: Helv. Phys. Acta **5**, 370 (1932)
5.12 L. Landau: Phys. Z. Sowjet Union **1**, 89 (1932); **2**, 46 (1932)
5.13 B.C. Eu: J. Chem. Phys. **55**, 5600 (1971)
5.14 B.C. Eu: J. Chem. Phys. **56**, 2507 (1972)
5.15 B.C. Eu, T.P. Tsien: Phys. Rev. A**1**, 648 (1973)
5.16 S.K. Kim: J. Math. Phys. **10**, 1225 (1969)
5.17 S.K. Kim: J. Math. Phys. **20**, 2153 (1979)

5.18 S.K. Kim: J. Math. Phys. **20**, 2159 (1979)
5.19 U.I. Cho, B.C. Eu: Mol. Phys. **32**, 1 (1976)
5.20 U.I. Cho, B.C. Eu: Mol. Phys. **32**, 18 (1976)
5.21 R.J. Cross, Jr.: J. Chem. Phys. **51**, 5163 (1969)
5.22 M.A. Wartell, R.J. Cross, Jr.: J. Chem. Phys. **55**, 4983 (1971)
5.23 M.E. Riley: Phys. Rev. A**8**, 742 (1973)
5.24 U.I. Cho, B.C. Eu: J. Chem. Phys. **61**, 1172 (1974)
5.25 U.I. Cho, B.C. Eu: Chem. Phys. Lett. **31**, 181 (1975)

Chapter 6

6.1 B.H. Bransden: *Atom Collision Theory* (Benjamin, New York 1970)
6.2 R.J. Glauber: *Lectures in Theoretical Physics*, Vol. 1, ed. by W.E.
    Brittin, L.G. Dunham (Wiley, New York 1959) p. 315
6.3 R.L. Sugar, R. Blankenbecler: Phys. Rev. **183**, 1387 (1969)
6.4 L. Wilets, S.J. Wallace: Phys. Rev. **169**, 84 (1968)
6.5 R.H.G. Reid, A.E. Kingston, M.J. Jamieson: J. Phys. B**10**, 55, 841 (1977)
6.6 G.B. Sorensen: J. Chem. Phys. **61**, 3340 (1974)
6.7 H.K. Shin: J. Chem. Phys. **41**, 2864 (1964)
6.8 E.E. Nikitin: *Theory of Elementary Atomic and Molecular Process in
    Gases* (Oxford University, Oxford 1974)
6.9 C.J. Joachain: *Quantum Collision Theory* (North-Holland, Amsterdam 1976)
6.10 B. Corrigall, K. Kuppers, R. Wallace: Phys. Rev. A**4**, 977 (1971)
6.11 R.P. Feynman: Rev. Mod. Phys. **20**, 367 (1948)
6.12 R.P. Feynman, A.R. Hibbs: *Quantum Mechanics and Path Integrals* (McGraw-
    Hill, New York 1965)
6.13 C. Morette: Phys. Rev. **81**, 848 (1951)
6.14 I.M. Gel'fand, A.M. Yaglom: J. Math. Phys. **1**, 48 (1960)
6.15 M.C. Gutzwiller: J. Math. Phys. **8**, 1979 (1967)
6.16 M.C. Gutzwiller: J. Math. Phys. **12**, 343 (1971)
6.17 C. Garrod: Rev. Mod. Phys. **38**, 483 (1966)
6.18 P. Pechukas: Phys. Rev. **181**, 174 (1969)
6.19 W.H. Miller: J. Chem. Phys. **53**, 1749 (1970)
6.20 S.A. Albeverio, R.J. Høegh-Krøhn: Inventiones Math. **40**, 59 (1977)
6.21 A. Truman: J. Math. Phys. **17**, 1852 (1976)
6.22 A. Truman: J. Math. Phys. **18**, 1499 (1977)
6.23 A. Truman: J. Math. Phys. **19**, 1742 (1978); **20**, 1832 (1979)
6.24 C. DeWitt-Morette, A. Maheshwari, B. Nelson: Phys. Rep. **50**, 255 (1979)
6.25 J.H. van Vleck: Proc. Nat. Acad. Sci. (USA) **14**, 178 (1928)
6.26 P.A.M. Dirac: *The Principles of Quantum Mechanics*, 4th ed. (Oxford
    University Press, Oxford 1958)
6.27 R. Schiller: Phys. Rev. **125**, 1100 (1962)
6.28 R. Schiller: Phys. Rev. **125**, 1109 (1962)
6.29 R.A. Marcus: J. Chem. Phys. **54**, 3965 (1971)
6.30 B. Leaf: J. Math. Phys. **10**, 1971 (1969)
6.31 B. Leaf: J. Math. Phys. **10**, 1980 (1969)
6.32 M. Moshinsky, T.H. Seligman: Ann. Phys. (NY) **114**, 243 (1978)
6.33 M. Moshinsky, T.H. Seligman: Ann. Phys. (NY) **120**, 402 (1979)
6.34 J. Deenen, M. Moshinsky, T.H. Seligman: Ann. Phys. (NY) **127**, 458 (1980)
6.35 J.R. Stine, R.A. Marcus: J. Chem. Phys. **59**, 5145 (1973)
6.36 J.N.L. Connor: Mol. Phys. **25**, 181 (1973)
6.37 J.N.L. Connor: Mol. Phys. **26**, 1271 (1973)
6.38 J.N.L. Connor: Discuss. Faraday Soc. **55**, 51 (1973)
6.39 J.N.L. Connor, H.R. Mayne: Mol. Phys. **37**, 1 (1979)
6.40 W. Güttinger: Prog. Theor. Phys. **13**, 612 (1955)
6.41 J.R. McDonald, M.K. Brachman: Rev. Mod. Phys. **28**, 393 (1956)
6.42 H. Goldstein: *Classical Mechanics* (Addison-Wesley, New York 1959)

6.43 R. Courant, D. Hilbert: *Methods of Mathematical Physics*, Vol. 2 (Inter-science, New York 1966)
6.44 T. Muir: *Theory of Determinants*, Vol. 2 (McMillan, London 1911)
6.45 R.H. Cameron, W.T. Martin: Trans. Amer. Math. Soc. **58**, 184 (1945)
6.46 C. Chester, B. Friedman, F. Ursell: Proc. Cambridge Philos. Soc. **53**, 599 (1957)
6.47 F. Ursell: Proc. Cambridge Philos. Soc. **61**, 113 (1965)
6.48 F. Ursell: Proc. Cambridge Philos. Soc. **72**, 49 (1972)
6.49 F. Nagel: Ark. Fys. **37**, 181 (1964)
6.50 N. Levinson: Bull. Amer. Math. Soc. **66**, 68 and 366 (1960)

Chapter 7

7.1 C. Zener: Proc. R. Soc. London Ser. A**137**, 696 (1932)
7.2 D.S.F. Crothers: Adv. Phys. **20**, 405 (1971)
7.3 B.C. Eu, T.P. Tsien: Chem. Phys. Lett. **17**, 256 (1972)
7.4 W.R. Thorson, J.B. Delos, S.A. Boorstein: Phys. Rev. A**4**, 1052 (1971)
7.5 J.B. Delos, W.R. Thorson: Phys. Rev. A**6**, 728 (1972)
7.6 M.S. Child: *Molecular Collision Theory* (Academic, New York 1974)
7.7 L.P. Kotova: Sov. Phys. JETP **28**, 719 (1969)
7.8 E.E. Nikitin, A.I. Reznikov: Phys. Rev. A**6**, 522 (1972)
7.9 R.E. Olson, F.T. Smith: Phys. Rev. A**3**, 1607 (1970)
7.10 U.I. Cho, B.C. Eu: J. Chem. Phys. **61**, 1172 (1974)
7.11 U.I. Cho: Ph.D. Thesis, McGill University (1976)
7.12 E. Merzbacher: *Quantum Mechanics*, 2nd ed. (Wiley, New York 1962)

Chapter 8

8.1 V.I. Osherov: Sov. Phys. JETP **22**, 804 (1966)
8.2 Y.N. Demkov: Dokl. Akad. Nauk SSSR **166**, 1076 (1966)
8.3 A.M. Woolley: Mol. Phys. **22**, 607 (1971)
8.4 J.S. Cohen: Phys. Rev. A**13**, 99 (1976)
8.5 M.L. Sink, A.D. Bandrauk: J. Chem. Phys. **66**, 5313 (1977)
8.6 B.C. Eu, N. Zaritsky: J. Chem. Phys. **70**, 4986 (1979)
8.7 U.I. Cho, B.C. Eu: Mol. Phys. **32**, 1 (1976)
8.8 U.I. Cho, B.C. Eu: Mol. Phys. **32**, 19 (1976)
8.9 U.I. Cho: Ph.D. Thesis, McGill University (1976)
8.10 G.V. Dubrovskii: Sov. Phys. JETP **31**, 577 (1970)

Chapter 9

9.1 G. Herzberg: *Molecular Spectra and Molecular Structure III. Electronic Spectra and Electronic Structure of Polyatomic Molecules* (Van Nostrand, New York 1966)
9.2 M.S. Child: J. Mol. Spectrosc. **33**, 487 (1970)
9.3 M.S. Child, R. Lefebvre: Chem. Phys. Lett. **55**, 213 (1978)
9.4 A.M.F. Lau: Phys. Rev. A**19**, 1117 (1979)
9.5 M.S. Child: Mol. Phys. **16**, 313 (1969)
9.6 A.D. Bandrauk, M.S. Child: Mol. Phys. **19**, 95 (1970)
9.7 M.S. Child, D.S. Ramsay: Mol. Phys. **22**, 263 (1971)
9.8 M.S. Child, C.E. Caplan: Mol. Phys. **23**, 249 (1972)
9.9 G.V. Dubrovskii, I. Fischer-Hjalmars: J. Phys. B**7**, 892 (1974)
9.10 B.C. Eu, N. Zaritsky: J. Chem. Phys. **69**, 1553 (1978)

Chapter 10

10.1 J.C. Tully, R.K. Preston: J. Chem. Phys. **55**, 562 (1971)
10.2 W.H. Miller, T.F. George: J. Chem. Phys. **56**, 5637 (1972)
10.3 T.F. George, Y.W. Lin: J. Chem. Phys. **60**, 2340 (1974)
10.4 J.R. Stine, J.T. Muckerman: Chem. Phys. Lett. **44**, 46 (1976)
10.5 J.T. Hwang, P. Pechukas: J. Chem. Phys. **65**, 1224 (1976); **67**, 4640 (1977)
10.6 J.D. Bjorken, H.S. Orbach: Phys. Rev. D**23**, 2243 (1981)

Chapter 11

11.1 H.C. Stier: Z. Phys. **76**, 439 (1932)
11.2 J.B. Fisk: Phys. Rev. **49**, 167 (1936)
11.3 S. Nagahara: J. Phys. Soc. Jpn. **9**, 52 (1954)
11.4 H.S. Massey, R.O. Ridley: Proc. Phys. Soc. London A**69**, 659 (1956)
11.5 P.M. Morse, H. Feshbach: *Methods of Theoretical Physics* (McGraw-Hill, New York 1953)
11.6 J.A. Stratton, P.M. Morse, L.J. Chu, J.D.C. Little, F.J. Corbató: *Spheroidal Wave Functions* (MIT, Cambridge 1956)
11.7 H. Bethe: *Quantum Mechanics of the Simplest Systems* (Russ. Transl. ONTI, 1935)
11.8 M.L. Sink, B.C. Eu: J. Chem. Phys. **78**, 4887 (1983)
11.9 M.L. Sink: Private communication
11.10 R. Schinke: Chem. Phys. **34**, 65 (1978)
11.11 B.C. Eu, M.L. Sink: J. Chem. Phys. **78**, 4896 (1983)

Author Index

Subject Index

Action 95
Adiabatic
  momentum 66,175
  potential 136
  representation 56,73
  state 137
  WKB wave function 52
Airy function(s) 10,12,13,37,38,59, 147
  Asymptotic formulas for 13
  differential equation 10,37,58
  Integral representation for 10, 99
  Wronskian of 38
Ambiguity
  group 82
  spins 82
Anti-Stokes lines 13,35,107
Arrangement channels 21
Asymptotic
  expansion 4,5,9
  evaluation 96
  formulas 47
  series 6
  solutions 7

Baker-Hausdorff formula 125
Bernoulli
  expansion 41,122
  number 40,203,204

Bessel functions
  spherical 182
Bohr-Sommerfeld quantization condition (rule) 45,146,197
Boundary condition 39,70,148

Cameron-Martin formula 94
Canonical
  commutation relations 83,85
  transformation 81,82,90
  transformation function 95
Channel
  Open 29
  Closed 29,30
Classical
  action integral 88,95
  deflection function 49
  S-matrix 93
  S-matrix theory 93
  turning point *see* Turning point
Clebsch-Gordan coefficients 26,28
Commutation relations 84
Complete integral 88
Complex trajectory method 165,173
Connection formulas 9,13,104
Cross section
  Differential 29,51
  Elastic scattering 50,185
  in prolate spheroidal coordinates 185,204,205

Molecular Collision Dynamics

Editor: J. M. Bowman
1983. 38 figures. XI, 158 pages. (Topics in Current Physics,
Volume 33). ISBN 3-540-12014-9

Contents: *J. M. Bowman:* Introduction. – *D. Secrest:* Inelastic
Vibrational and Rotational Quantum Collisions. – *G. C. Schatz:*
Quasiclassical Trajectory Studies of State to State Collisional
Energy Transfer in Polyatomic Molecules. – *R. Schinke,*
J. M. Bowman: Rotational Rainbows in Atom-Diatom Scattering.
– *M. Baer:* Quantum Mechanical Treatment of Electronic
Transitions in Atom-Molecule Collisions. – Subject Index.

S. Büttgenbach
Hyperfine Structure
in 4d- and 5d-Shell Atoms

1982. 14 figures. VIII, 97 pages. (Springer Tracts in Modern
Physics, Volume 96). ISBN 3-540-11740-7

Contents: Introduction. – Theoretical Considerations. – Experi-
mental Methods. – Experimental Hyperfine Interaction Con-
stants for 4d- and 5d-Shell Atoms. – Intermediate Coupling
Wavefunctions, Atomic g_J Values and Parametric Interpretation
of Level Isotope Shifts. – Effective Radial Parameters of the
Magnetic Dipole and Electric Quadrupole Interaction. – Nuclear
Moments and Hyperfine Anomalies. – Concluding Remarks. –
References. – Abbreviations in this Work. – Subject Index.

Laser Spectroscopy of Solids

Editors: W. M. Yen, P. M. Selzer
1981. 117 figures. XI, 310 pages. (Topics in Applied Physics,
Volume 49). ISBN 3-540-10638-3

Contents: *G. F. Imbusch, R. Kopelman:* Optical Spectroscopy of
Electronic Centers in Solids. – *T. Holstein, S. K. Lyo, R. Orbach:*
Excitation Transfer in Disordered Systems. – *D. L. Huber:* Dyna-
mics of Incoherent Transfer. – *P. M. Selzer:* General Techniques
and Experimental Methods in Laser Spectroscopy of Solids. –
W. M. Yen, P. M. Selzer: High Resolution Laser Spectroscopy of
Ions in Crystals. – *M. J. Weber:* Laser Excited Fluorescence
Spectroscopy in Glass. – *A. H. Francis, R. Kopelman:* Excitation
Dynamics in Molecular Solids.

Radiationless Processes in Molecules and
Condensed Phases

Editor: F. K. Fong
With contributions by *D. J. Diestler, F. K. Fong, K. F. Freed,*
R. Kopelman, J. C. Wright
1976. 67 figures. XIII, 360 pages. (Topics in Applied Physics,
Volume 15). ISBN 3-540-07830-4

Contents: Introduction. – Energy Dependence of Electronic Re-
laxation Processes in Polyatomic Molecules. – Vibrational Re-
laxation of Molecules in Condensed Media. – Up-Conversion
and Excited State Energy Transfer in Rare-Earth Doped Mate-
rials. – Exciton Percolation in Molecular Alloys and Aggregates.

Springer-Verlag
Berlin
Heidelberg
New York
Tokyo

Applied Physics B

Photophysics and Laser Chemistry

Fields and Editors:

Laser Physics and Spectroscopy

High-Resolution Laser Spectroscopy:
V. P. Chebotayev, Novosibirsk
Laser Spectroscopy: **T. W. Hänsch,** Stanford U.
Quantum Electronics: **A. Javan,** MIT
Ultrafast Phenomena: **W. Kaiser,** TU München
Laser Physics and Applications:
H. Walther, U. München

Chemistry with Lasers

Chemical Dynamics and Structure: **K. L. Kompa,** MPI
Garching
Laser-Induced Processes: **V. S. Letokhov,** Moscow
Dye Laser and Photophysical Chemistry:
F. P. Schäfer, MPI Göttingen
Laser Chemistry: **R. N. Zare,** Stanford U.

Photophysics

Optics: **W. T. Welford,** Imperial College
Nonlinear Optics and Nonlinear Spectroscopy:
T. Yajima, Tokyo U.

Editor: **H. K. V. Lotsch,**

Springer-Verlag, P. O. Box 105280,
D-6900 Heidelberg 1, Federal Republic of Germany

Special Features:
● rapid publication (3–4 months)
● no page charges for concise reports
● 50 offprints free of charge

Springer-Verlag
Berlin
Heidelberg
New York
Tokyo

Subscription information and/or **sample** copies are
available from your bookseller or directly from
Springer-Verlag, Journal Promotion Dept.,
P. O. Box 105280, D-6900 Heidelberg, FRG